D1261255

THE MANIPULATION OF LIFE

OTHER NOBEL CONFERENCE BOOKS AVAILABLE

FROM HARPER & ROW

Nobel Conference XVII, *Mind in Nature*
Nobel Conference XVIII, *Darwin's Legacy*

NOBEL CONFERENCE XIX

Gustavus Adolphus College, St. Peter, Minnesota

The Manipulation of Life

Edited by Robert Esbjornson

With Contributions by Lewis Thomas, Karen Lebacqz, Christian Anfinsen, Willard Gaylin, Clifford Grobstein, and June Goodfield

Harper & Row, Publishers, San Francisco

Cambridge, Hagerstown, New York, Philadelphia
London, Mexico City, São Paulo, Singapore, Sydney

To Richard Q. Elvee,

who personifies the Nobel Conversations about science, religion, and the arts.

 For information address Harper & Row, Publishers, Inc., 10 East 53rd Street, New York, NY 10022. Published simultaneously in Canada by Fitzhenry & Whiteside, Limited, Toronto.

FIRST EDITION

Designer: Jim Mennick

Library of Congress Cataloging in Publication Data

Nobel Conference (19th : 1983 : Gustavus Adolphus College)
THE MANIPULATION OF LIFE

1. Genetic engineering—Moral and ethical aspects—Congresses. 2. Medical ethics—Congresses. 3. Bioengineering—Moral and ethical aspects—Congresses.
I. Esbjornson, Robert. II. Thomas, Lewis, 1913– III. Title.
QH442.N63 1983 174'.2 84–47723
ISBN 0-06-250296-4

84 85 86 87 88 10 9 8 7 6 5 4 3 2 1

Contents

Preface:
The Making of Nobel XIX

The essays in this book came out of the nineteenth Nobel Conference at Gustavus Adolphus College, St. Peter, Minnesota, October 4–5, 1983. The conferences are on science-related topics and were created as a living memorial to Alfred Nobel, with the permission and endorsement of the Nobel Foundation in Stockholm. The idea for the conferences came from Nobel Laureate Glenn Seaborg, and over the years more than one hundred scholars, including twenty-two Nobel Laureates, have presented lectures at the conferences.

What makes a Nobel conference? An interested audience, attracted by a significant topic, presented by competent and interesting persons who have devoted years of their lives to the subject—and a lot of hard work in preparation for the events of the conference.

Over five thousand people were attracted to Nobel Conference XIX. Their interest was manifest. After each lecture they sent up more questions for discussion than the panel could handle. Clusters of students from high schools and colleges and others gathered around the speakers at the end of the sessions.

The significance of the topic has a great deal to do with generating that interest. Advances in scientific knowledge and their applications in medical practice touch human lives at most intimate and critical times. There is a widespread concern about the escalating costs that these advances have brought, and many are beginning to ask how we can contain them. The topic's significance is magnified because it seems humans now have to face the prospect of directly engaging themselves in altering boundaries of human power that only a few years ago were believed to be fixed.

The idea for the topic of Nobel XIX was generated in conversations and letters between Richard DeRemee, M.D., internal medicine specialist at Mayo Clinic in Rochester, Minnesota, and Nobel Conference Director Richard Elvee, College Chaplain. DeRemee proposed that the 1983 conference be on the social responsibilities of science and society's responsibilities for science, with a focus on medical research and practice. As planning proceeded it seemed sensible to concentrate on advances in the fields of genetics, molecular biology, developmental biology, and related sciences—those sciences the applications of which could affect the formation and beginning of life. Not only do these developments have important consequences for the shaping of life, but they have become the base for a new and growing industry attracting hundreds of millions of dollars for investment.

The planning committee chose as participants people well known within the scientific and medical communities for their knowledge and experience and ability to communicate effectively with the general public. It is gratifying to have our judgment verified by their lively and lucid presentations at the conference sessions. The college is grateful to them for their generous and amiable participation at every stage of the conference, and for the provocative essays they provided for this volume. As chairman I can declare that presiding at this conference was one of the most enjoyable assignments I have ever had.

I must also commend and thank the many people whose cooperation, support, and work were essential if the conference was to be kept from collapsing into chaos: the members of the planning committee; Elaine Brostrom and Steve Waldhauser and the public relations staff; Linda Miller and food service staff; and Dale Haack and Dennis Paschke and their crews, who provided the physical facilities for the conference. Because of the careful attention to detail, alert support during the conference, and hard work, it was an easy conference to organize and moderate.

The Lund family, whose generous grant has made these annual conferences possible, are worthy of special thanks and respect.

Introduction: Early Morning Thoughts of a Chairman After a Conference

They are gone—all of them, the speakers, the audience. The empty chairs and discarded programs have been cleared away. An empty arena and silence are mute testimony to a significant event that has passed into several thousand memory banks. What was it all about? What remains after reflection? Nobel Conference XIX was about the human responsibilities that come along with the recent medical advances that give humans greater power to manipulate life.

About Manipulating Life . . .

Manipulating is not a novel human activity and not necessarily nefarious. The word is of Latin origin—from *manus*, a hand; and *pellere*, to push; or *plere*, to fill. It means literally "to get things into one's hands," or to get matters in hand and under control. It can refer to efforts to influence the biological or psychological nature or functions of humans and to shape their environments in beneficial directions. Humans have manipulated objects and environments ever since they became a distinct and intelligent species; they have manipulated one another for various personal and social purposes through myth and ritual, charismatic and powerful heroes and leaders, education and ideology. And they have used various medical therapies to alleviate pain, control illness, and prolong life. The capacity to "take matters in hand"—to manipulate—seems inherent in our humanity.

What is so novel about our current manipulative capacities that we are inclined to regard them as "revolutionary"? And why has the word taken on an ominous connotation? Manipulation is a technical business, and our capacities have been sharply enhanced by advances in biomedical research and practice. But it is also a *moral* affair involving humans who may risk being harmed or who may enjoy potential benefits. The conference lectures and conversations explored the new advances of recent years and the human responsibilities that have come along with them.

About Medical Advances . . .

Recent advances in biomedical science and practice have been called a revolution. A revolution is a turning in a new direction. The word implies basic changes in underlying ideas, not just cosmetic alterations in forms, rules, and controlling parties.

Medical practice was not significantly linked to the scientific method until the nineteenth century, when speculative notions about the causes of illness were replaced by knowledge that had been accumulated by careful observation, experimentation, and reasoning. Such innovations as the stethoscope, anesthesia, aseptic procedures, and vaccines began to change the practice of medicine.

In the keynote essay of the conference, "The Limitations of Medicine as a Science," Lewis Thomas points out that until the late 1930s the functions of the doctor were primarily to make an accurate diagnosis, explain the likely prognosis of a disease or injury, and provide care and comfort. There were only a few tools for therapeutic medicine, such as insulin for diabetes, liver extract for pernicious anemia, and vitamin B for pellagra. In the late 1930s two great advances in medical treatment occurred that shifted the focus from care to cure and from diagnosis to the repair of the ravages of illness and injury. They were the discovery and application of antibiotics to cure infectious disease and the development of vastly improved surgical techniques. These revolutionary advances have left the impression that there are now no diseases beyond treatment and cure. Thomas presents a formidable list of major problems for human health in order to show that medicine still has a long way to go. He calls attention to the expensive "half-way technologies"—such as organ transplants, chronic renal dialysis, and the artificial heart—that only alleviate symptoms and provide only limited prolongation of life.

The real advances of recent years have been in basic research aimed at the underlying mechanisms of disease. Thomas describes the exhilarating optimism about research among immunologists, experimental pathologists, developmental biologists, and biochemists—and on the frontiers, the molecular biologists and geneticists. Such scientists are "in possession of research techniques that permit them to ask, and then answer, almost any question that pops into their minds."

The essays focus on the actual and prospective problems and possibilities of advances in those fields that could bring about changes in reproductive technologies, invention of new forms of life, gene therapy, and the more speculative but not entirely impossible power to alter germ cells of humans so as to change human nature and behavior. Pushed to the periphery are the more immediate problems that surface in the day-by-day work of the physician. Such questions as when to start or stop life support in intensive care units, when and how to tell a patient the results of a diagnosis, what reforms of the medical care system should be advocated, and how to deal with the declining public trust in doctors that has accompanied the rise in expectations and dissatisfaction with medical care are not the focus of the essays.

There are significant reasons for giving attention to issues related to reproductive and genetic matters. Thirty years have passed since Watson and Crick cracked the genetic code in 1953, and it was ten years ago that a new laboratory technique called recombinant DNA was discovered. New prospects in basic research and applications have opened up a novel range of concerns. Since the birth of Louise Brown in England in 1978 it is estimated that well over one hundred and fifty babies have been born as a result of a procedure of external or *in vitro* fertilization. It is already possible to diagnose several genetic diseases by means of sophisticated techniques such as amniocentesis, sonography, and chorionic villi biopsy. Human insulin, interferon, and natural growth hormone have been manufactured in laboratories using recombinant DNA and are on their way to production and distribution. In 1980 the U.S. Supreme Court authorized the patenting of new life forms. Genetics industries attracting investments of hundreds of millions of dollars are in the process of developing industrial, agricultural, environmental, and medical applications of the new techniques. Lewis Thomas gives considerable attention to the prospects of advances in eliminating pathological aging, a major cause of personal distress and social costs. Christian Anfinsen refers to the recent production in laboratories of "giant mice" by altering the germ cells and foresees rapid development of similar experiments on other animals and, perhaps within twenty years, on humans; he stresses the potential applications of biotechnology to improve food resources and population controls. Clifford

Grobstein focuses on external human fertilization, gene transfer, and a possible combination of the two that could result in the first direct human intervention in human hereditary development on an intergenerational scale.

The conference took place in the context of much interest among the general public generated by the growing knowledge of genetic diseases. In the early months of 1983 there was a rash of publicity about the efforts of Jeremy Rifkin in his book *Algeny* and of an unofficial group of religious leaders and scientists who advocated limiting the use of recombinant DNA to therapeutic applications. The President's Commission for the Study of Ethical Problems in Medicine and Biomedical and Behavioral Research released its study on genetics, entitled *Splicing Life*, some months earlier. The National Council of Churches had issued a report, *Human Life and the New Genetics*, in 1980.

It seems appropriate, in retrospect, that the conference should focus as it did on manipulating life. It was an anniversary marking significant advances in biomedical science and it took place in a climate of public interest and concern.

About Human Responsibility . . .

Medical advances are not just scientific and technological developments but also moral events, because they represent humans with inescapable decisions about what we ought to know and do. New knowledge extends power, and new duties accompany these powers. As long as we are subject to the fate of random events beyond our control, we can excuse ourselves, but when we are able to do something to help (or harm) ourselves and others, we are responsible for the consequences and vulnerable to stress and anxiety. Not only powerless and oppressed people who suffer deprivation have concerns about what to do.

People in power also live a precarious existence, because they are uncertain about their capacity to control events or properly use their power. They know from experience that humans can become slaves rather than masters of the very powers they have created. Not using the powers we have can be as irresponsible as abusing them. New powers are complex moral burdens. They increase control at the same time they increase the risks of losing control. The extension of benefits increases the expectation of receiving more of them, and these very extensions and expectations are forces that destabilize the social mechanisms of distribution already in place.

What is so novel about our current manipulative capacities that they have caused moral perplexity and aroused the concerns of so many? We are beginning to manipulate biological processes more directly than ever before. These advances portend future capacities to control life processes, create new forms of life, and even to change human nature and behavior. Although there is some disagreement about what could be done through genetic engineering, external fertilization, and other manipulative technologies, the writers are generally cautious and conservative, particularly with reference to what could and should be done about manipulating germ cells, which would affect future generations. There is a strong assertion from the scientists that scientific research, powered by curiosity, is hard to control and its breakthroughs are nearly impossible to predict. Clifford Grobstein stated the issue in the introduction to his book *From Chance to Purpose*:

> The recent sharply rising capability for deliberate and direct intervention in hereditary and reproductive processes introduces a new dimension. Options are now on the horizon that may incorporate human purpose *directly* into human progression. In the past we made many choices without awareness or concern about the hereditary and evolutionary consequences.

Today certain choices are being presented for which we can no longer shift responsibility, whether to Divinity, Chance, or Unkind Fate. We are acquiring options to intervene deliberately and purposefully in fundamental matters that hitherto were beyond our control. Having long been an environment-creating species, we now face the issue of becoming a self-creating species. This is the nature of the new great transition we are approaching.

It is terribly tempting to hide our heads under the blankets so we will not see what Karen Lebacqz calls "the ghosts on the wall." People, not the least scientists and doctors, tend to avoid talking about the moral questions they encounter in their lives, but there is no escape.

We cannot escape because "the ghosts" are there. In her essay, "The Ghosts are on the Wall: A Parable for Manipulating Life," Lebacqz lists new techniques for manipulating life that generate unresolved legal questions, troubling ethical dilemmas, persistent technical problems, and uneasiness about threats to cherished values. The "ghosts" are bothersome issues accompanying these developments. She uses the Frankenstein image to express common fears about unleashing a monster beyond human control.

A second reason why we cannot escape moral questions is that humans are inherently moral agents. This is the presumption in the discussions about human responsibility. We may wish for a "morality shelter" as safe as our tax shelters so we can postpone responsible action, but we wish in vain. Evasion is not an escape from choice; it is an irresponsible choice, because it leaves options and consequences unreviewed. We are moral agents because we make choices and we care. It is our nature to do so.

Humans can make choices because they can transcend their biological drives enough to study and direct them and their sociological conditioning enough to criticize and change them. In his essay, "What's So Special About

Being Human?", Willard Gaylin describes the unique traits of mind and heart that makes us different from the rest of life. Although we are connected to and interactive in the web of all living organisms, we are able to manipulate ourselves and our environments so as to challenge and change them. Although the capacity to choose is grounded in biological and social conditions, it is not bound by them.

There is something *strange* about humans. We seem ever to be strangers in our own homes, lured by an invading vision that inspires new explorations, ideas, stories, and inventions. Lewis Thomas alludes to this when he mentions the attraction of basic research, which he calls "the utter strangeness of nature." Karen Lebacqz's essay and comments imply this, because she maintains that the scientific model of empirical and rational thought is not enough. There is something more. There are stories and perhaps a larger story that includes all others and embraces the diversity of human experience. She calls for opening our minds to new paradigms, which she defines as "deeply-held modes of thought that structure our ways of thinking"; ways that are more responsive to intuitive, imaginative, and playful kinds of thinking, to nurturing modes of being, and to the voices of the oppressed.

We care because humans must nurture and protect their young during a long period of biological and social dependency. In her essay, "Without Laws, Oaths or Revolutions," June Goodfield emphasizes caring as the foundation of responsibility. Although it is grounded in the necessity of caring for the young during the long period of biological and social dependency, it is not confined by its grounding. It becomes a matter of moral experience when we make decisions about caring, decisions about *what* is good, beneficial, and valuable; and decisions about *whose* good we care about. What medical advances are beneficial and for whom are issues that pervade several conversations. The basic issue is the question of community—who should belong, who should be "on the inside," who should benefit from medical advances, who should make the decisions about what quality of life is and about the distribution of costs and benefits. Is there anyone trustworthy enough to be the elite corps of decision makers? Is it right to invest thousands, even millions of dollars in expensive high medical technology available only to a few well-to-do people while thousands and millions of people do not have access to basic maintenance or adequate nutrition? What should we do about the deprivations of people in other countries whose land is used to grow coffee, bananas, marijuana, and poppies to satisfy the appetites of affluent people in developed countries at the expense of the basic needs of indigenous populations? Is it right to support the goals and projects of scientists and technicians who are developing the expensive medical and military technology of our society while we deny adequate support to and opportunities for people in education and human services?

The question *For whom should we care* is inescapable. The answer that comes through again and again is *for all*. It is justice that requires sharing and distributing benefits and costs widely and as fairly as we can. The commitment to global inclusiveness is apparent in Christian Anfinsen's concern about food supplies and health in Third World countries, in Lewis Thomas's proposal for an international medical corps comparable to the Peace Corps, in June Goodfield's international involvements and concerns, in Karen Lebacqz's proposal that the voice of the poor and oppressed be considered in research and decision-making models, and in Clifford Grobstein's proposals to reform policy-making procedures to be inclusive enough to involve all affected parties.

However, making advances in justice is far from easy. Each of us places unilateral demands on the resources of the natural environment and on the social, economic, and medical systems and thus puts stress on their carrying capacities. If the pressures become excessive, the systems are damaged and all of us are hurt. Even if this is so, the personal loss to each party of limiting his or her demands is such that each of us is unwilling to pay the price of self-limitation. Unrestricted individual gain, whether for persons or special interest groups, results in eventual damage to the sustaining systems. This is the prospect that worries Lewis Thomas and others who see the costs of medical care escalate and who know that they must not be allowed to do so. Must we choose—or are we already choosing—an elitist solution by providing expensive care only for those who can afford it and a trickling-down of the crumbs of minimal care to the poor? Are we ordinary mortals controlled by extraordinary people such as scientists, physicians, or generals, whose knowledge and power are beyond us? Is our sense of autonomy and self-respect undermined by their intrusive technology, which is so elaborate and complex that we are awed into passivity? We trust, and distrust, scientists and doctors. We hardly trust ourselves, because of our diminishing ability to understand and control the complexities of our system. How can trust, so necessary for community and cooperation and caring properly for one another, be strengthened? Are we creating policy-making and regulatory procedures adequate for establishing the direction and control of medical advantages so that their costs and benefits are distributed fairly?

In his essay, "Manipulating Life: The God-Satan Ratio," Clifford Grobstein asserts, "We are in a period of improvisation and invention in mechanisms of policy formation to match the thrust of our scientific and technological advances," and he outlines further steps of a new approach that are needed. He calls for the widest possible participation in the process and proposes three ethical principles that should provide the basis for the formation of new policies.

Why is there a "cry" for justice? Why are we moved by the cry? Why does justice have the power to lay hold of our efforts to improve our caring for one another? Why should we care about anyone but ourselves and those who are the primary exchangers in our lives, the people close to us? Why should we be generous and fair in our caring? What is the lure or prod that makes us disconcerted and ashamed by injustice and our self-centeredness? Survival? We are at least smart enough to see that injustice and selfishness are contrary modes that threaten our safety. We can see that self-interest is destructive when it results in excluding any who need care. However, survival is not our only concern. How can we account for sacrifice, for the real presence of persons among us who sacrifice benefits, comforts, and even their lives for others, even though they have no expectations of gain for themselves? Is it because they hope their sacrifice will contribute to the survival of the species? Perhaps, but "species" is an abstraction. For future children? Future children are only possibilities, not actualities. There is *something more*—what Daniel Maguire calls "our mysterious tendency to esteem certain person-related values so highly that when they are at issue, we will die for them. Or if we do not have the courage or the opportunity to do so, we will admire those who do and will call them heroes" (Maguire, *The Mortal Choice*, 74). This strange propensity presents a fundamental question, which is essentially a religious question: what or who is the strongest and most splendid reality in our lives, luring us toward becoming, caring and sharing? Manipulation of life, scientific research, and medical advances are more than scientific, technical, and moral affairs. They present a theological question: Is human creativity expressed in science and technology a violation of some kind of natural or divine order?

The question of whether humans are assuming godlike prerogatives is discussed in the essays. The term "playing God" is frequently used to refer to the extension of human knowledge and control into hitherto uncharted territory considered to be God's turf. Nowhere is this question more urgent and poignant than in decisions about genetic manipulation and external fertilization, which could give humans the power to create life. Grobstein addresses this question directly: "The advent of this potential new age of human intervention seemed close to the ultimate in arrogant encroachment on divine prerogative." However, he challenges the wisdom of regarding humans as agents of either God or Satan: "Do we not strike closer to human responsibilities by facing as our own the choices between moral alternatives?" He asserts that "our growing options and powers have increased to levels that force attention to human responsibilities that were once assigned to the gods." As a theological ethicist, Karen Lebacqz agrees that the issue is wrongly conceived when it is posed in terms of "playing God." In the Judeo-Christian tradition, neither knowledge nor control is intrinsically wrong. For her, a more basic question is whether humans could challenge the sovereignty of God or call God into question. The "monster" is our propensity to ask the wrong questions, to confine our models of thought and action too narrowly. The story is larger than our imaginations can conceive, so we must have a "radical scepticism" about all our knowledge and power.

There is little agreement about what or who is this "foreign force" that invades our worlds but yet can elicit responses of creating new values and caring for others. Even within a single denomination of Christians there is significant diversity of theological beliefs. We are confronted by a mystery we can know only by symbol and story and not directly by reason and empirical evidence. Karen Lebacqz focuses attention on this issue by her provocative challenge to the predominant scientific paradigm of Western society, which tends to define what reality is by empirical evidence and logical thinking. There are other realities that come through other sources such as fantasy, imagination, parable, and the poignant cries of the oppressed. There is an "otherness" about our experience that evokes rich and varied responses. Lewis Thomas calls attention to the emergence of new diseases, which pose new challenges and force us to be on the move, even though we do not want to change and would rather settle down.

Is there some reason for our resistance to assuming responsibility for caring, for change? June Goodfield sees the resistance in our persistent pattern of turning in on ourselves, a traditional way of describing sin. Our sin is the tendency to claim benefits for ourselves and to secure our own gains and to turn away from others. This tendency is a violation of the very obvious reality of human interdependence with other humans and with the biological and physical worlds in which we live. Sin violates the very nature of humans to care, which Goodfield emphasizes in her essay and which is unanimously acknowledged in the last conversation. There is no support from writers for the view that humans sin by their search for knowledge or by their possession and use of increasing powers. Responsibilities have to do with careful research, cautious application and sensitivity to all the people who could benefit or suffer from the advances in research and medical practice.

In the conversation relating to Grobstein's essay, Karen Lebacqz proposes a principle that requires a covenant between those who do research and those who would be the recipients of the effects of that research. Research subjects and patients should work *with* and not *under* scientists and physicians, because they have responsibilities also for seeking and telling the truth and for healing. Paul Ramsey in his book *Patient as Person* makes covenant a key image for responsibility. The word "responsibility" comes from two Latin words—*re*, back to; and *spondere*, to pledge. Covenants involve making and keeping promises, and at the core of covenant is mutual promise. The word "promise" is from two Latin words—*pro* and *mittere*—to send forth. Promise is a commitment to future action. Together the words responsibility and promise suggest that moral life is not only a matter of faithfulness but hopefulness, not only of looking back but also of moving forward.

Both science and ethics are future-oriented. Although they are disciplines that draw upon accumulated knowledge and wisdom, they are enterprises that look ahead to "what next." We know only a tiny bit about nature, Thomas says in his essay, and we want to know more, so science presses forward in its basic research. We want to improve medical care, so physicians continue to seek new applications of expanding knowledge. Ethicists are primarily concerned about the question of what we ought to do, which is a question about future action. Their discipline is not primarily prescriptive, settling moral issues ahead of time; nor is it preventive, forestalling moral difficulties. But it is a preparatory discipline helping people to get ready for decisions that must be made as new advances come to them. All humans are ethicists when they consciously pursue moral values. It is the task of professional ethicists, says Daniel Maguire, "to attempt to bring sensitivity, reflection, and method to the way in which his fellow humans have learned to do, or stumbled into doing, ethics. The ethicist stops to think how moral judgments are and should be arrived at; most people do not stop to think. They should have something to learn from the ethicist who does" (*Death by Choice*, 77).

Nobel Conference XIX was an attempt to look ahead and not just to assess past attainments, so that we might be better prepared for the critical choices that we cannot escape because they come with the changes wrought by new knowledge and power. Clifford Grobstein concludes his essay with a significant statement along this line: "Our human thrust, as a community, is to move outward to a wider existence—physically, intellectually and spiritually. Exactly what this means will not be fully defined except as it is achieved and then it will only be a platform for a further step toward what lies beyond. For better or for worse, for good or for evil, whether divine or satanic, that is what it means to be human."

Contributors

LEWIS THOMAS
Chancellor of Memorial Sloan-Kettering Cancer Center, New York, and Professor of Pathology and Medicine at Cornell University Medical College; member of the President's Scientific Advisory Committee (1967–70); recipient of Woodrow Wilson Award from Princeton University (1980) and the Kober Medal from the Association of American Physicians (1983); author of *The Lives of a Cell* (1974), *The Medusa and the Snail* (1979), *The Youngest Science: Notes of a Medicine-Watcher* (1983), and *Late Night Thoughts on Listening to Mahler's Ninth Symphony* (1983).

KAREN LEBACQZ
Professor of Christian Ethics at the Pacific School of Religion; Hastings Center fellow (1973–82); author of *Professional Paradox: Morality in Ministry* (1984).

CHRISTIAN ANFINSEN
Professor of Biology, Johns Hopkins University; 1972 Nobel Prize in Chemistry; President, American Society of Biological Chemists (1971–72); member of the Council of the National Academy of Sciences (1974–77), and of the Pontifical Academy of Science (1981); author of *The Molecular Basis of Evolution* (1959).

WILLARD GAYLIN
President of Hastings Center (Institute of Society, Life

Sciences, and Ethics) and Training and Supervising Psychoanalyst at the Columbia Psychoanalytic Center; winner of the Van Gieson Award for outstanding contributions to the mental health sciences; author of *In the Service of Their Country* (1970), *Caring* (1976), *Doing Good* (1978), *Feelings* (1979), and *The Killing of Bonnie Garland* (1982).

CLIFFORD GROBSTEIN
Professor of Biological Sciences and Public Policy at the University of California at San Diego; winner of the Brachet Award from the Belgian Royal Academy of Science in 1959; past president of the American Society of Zoologists and the Society for Development and Growth; author of *Strategy of Life* (1965), *A Double Image of the Double Helix* (1979), and *From Chance to Purpose* (1981).

JUNE GOODFIELD
Associate scientist and adjunct member of Memorial Sloan-Kettering Cancer Center; Phi Beta Kappa lecturer for Association for the Advancement of Science; author of *The Siege of Cancer* (1975), *Playing God* (1977), *Reflections on Science and the Media* (1981), and *An Imagined World* (1981).

The Limitations of Medicine as a Science

LEWIS THOMAS

Fifty years ago seems a short distance in the records of science and technology; no time at all in the millennia of the long history of medicine. Doctors can trace their roots, doctrine by doctrine, back through the Roman Empire into Ancient Greece past Aesculapius, all the way back to the shamanism of the Indo-europeans, and no doubt further still into the earliest equivalents of witch-doctors conjuring out the illness of the first hunters and gatherers. It is a long professional lineage, doctor after doctor. We have the oldest roots for our occupational labels, still resonating inside our titles: "Doctor" from Indo-european *dek*, meaning "proper" and "acceptable," with cousin words like "orthodox" and "dogma"; "Physican" from *bheu*, meaning "nature," becoming *phusis* in Greek, then "physic" and "physics"; "Leech" from *leg*, meaning "to collect and speak," turning later into "logic" and "legend." (Leech, the worm, came along later by a process of assimilation, from more obscure origins, but still meaning "collecting," like the doctor, and used for centuries in medicine, signifying measuring out and taking appropriate measures.)

All our philological instructions are still with us, exhorting the medical profession to behave properly and with decorum and decency, to go carefully with nature, to provide sound and measured advice. The word "medicine" has within it, as etymological reminders, cognate words like "moderate" and "modern," but also "modest."

This is as it should have been but, alas, has never been. The history of medicine has never been a popular subject for study in the medical schools, partly because it has been for most of its stretch of time so embarrassing. Michel Eyquem de Montaigne, in his essay *Children and Fathers*, lashed out at the profession and its pretensions. "The doctors," wrote Montaigne, "are not content with having control over sickness; they make health itself sick, in order to prevent people from being able at any time to escape their authority. . . . I am not upset at being without a doctor, without an apothecary, and without help, from which I see most people more afflicted than by the disease."

Montaigne could serve as the earliest standard bearer for today's proponents of preventive medicine, even for those activists pressing for a new "holistic" medicine and what are called "alternative" methods of health care. He wrote, "We disturb and arouse a disease by attacking it head on. It is by our mode of life that we should weaken it, by gentle degrees . . . the drug is an untrustworthy assistant, by its nature an enemy to our health."

The principal pharmacopeia at medicine's disposal, and probably the one that so infuriated Montaigne, was the ancient collection of drugs employed for violent purgation. Purging and, later, bleeding were the two things that doctors did for serious illnesses, and they continued to place fervent reliance on these clear up through the nineteenth century. George Washington is reported to have died, hale and hearty at the age of sixty-seven, after being bled for treatment of a quinsy sore throat; two and a half quarts of his blood were removed, probably bringing him into shock, within an inch of his life, and then perhaps beyond. It was not until the late nineteenth century that medicine's ancient motto, *primum non nocere* (first do no harm), was taken seriously and literally. At that time, most of medicine's science and therapeutic technology were abandoned; Sir William Osler and his school initiated what came to be called "therapeutic nihilism"; doctors learned to confine their scholarly efforts to studies of the natural history and pathology of disease; only a few remedies survived, such as digitalis, morphine, quinine, and aspirin. The chief functions of the doctor were threefold: To make an accurate diagnosis, to explain to the patient and family how the illness would most likely turn out, and then to stand by, making sure that good nursing care, nutrition, and comfort were at hand. This was essentially the kind of medicine that I was taught at Harvard Medical School in the mid-1930s, and none of us—students, faculty or patients—had any notion that it would ever be any different. By that time we possessed insulin for diabetes, liver extract for pernicious anemia, and vitamin B for pellagra; we regarded these as miraculous anomalies, pure gifts from heaven, but we did not acknowledge them as the fruits of science, for we did not really think of therapeutic medicine as any kind of science.

Then, in the late 1930s, sulfanilamide appeared, followed within a few years by penicillin and then the other antibiotics, and medicine was transformed overnight, so to speak, from an ancient art to a modern science—or so we came to believe. We could henceforth give up, we thought, the merely supportive function of standing by. Reassurance would no longer be our principal occupational service. We could abandon all placebos. Montaigne could be exorcised, Plato as well; Plato had hurt our professional feelings long ago by remarking that "it is for doctors alone to lie in all freedom, since our safety depends on the vanity and falsity of their promises." We were, after the conquest of infectious disease that began a half-century ago, home and dry. Or so we thought.

Almost simultaneously with the emergence of antimicrobial therapy—and in part because of it—surgery began a comparable revolution. In the years since, surgical techniques, associated with vastly improved methods for maintaining fluid balance, blood volume, and oxygenation, advanced to a level of sophistication and power that was inconceivable before. Organ transplantations, open heart surgery, the repair of minute blood vessels, the replacement of severed limbs, and extensive procedures for the removal of previously unapproachable cancers became everyday, routine procedures.

These, I think, have been the two great advances in medical treatment. They are indeed wonderful, but they leave in the public mind the impression that medical care has come along so fast and so far that there are now no diseases beyond the reach of treatment and cure. It is, of course, not so. Still on medicine's agenda of essentially unsolved, untreatable, and unpreventable diseases is a long roster of major problems for human health: coronary heart disease, stroke, at least 50 percent of cancers, the vascular complications of diabetes, rheumatoid and degenerative arthritis, multiple sclerosis, schizophrenia, the senile dementias, cirrhosis, emphysema, nephritis, and a long list in addition of less common but disabling and potentially lethal disorders.

Looked at in this way, medicine has a great part of its full distance still to go. While very much better than the medicine of fifty years ago, it owes much of its eminence, to paraphrase Gibbon, to the flatness of the preceding terrain.

The real advances, celebrated by the biomedical sciences during recent years, have been in basic research aimed at what will almost undoubtedly turn out to be the important underlying mechanisms of disease. The new information is exciting, even breathtaking in its implications, but it has to go farther and much deeper before it will open up new approaches to therapy. A recent and exceptional example of what is needed can be seen in the treatment of hypertension; a new and rational pharmacology, aimed at intervening at a fundamental level of enzymology, is providing tailored drugs for the control of malignant hypertension. The high hope in medicine is that as more is learned about the working parts of all the other diseases on that list, equally effective remedies will be devised.

But it has not happened yet, and there is an enormous amount to be learned. Meanwhile, we are stuck with a level of technology for managing some of the commonest diseases that is neither decisive nor conclusive, is enormously expensive, and while it provides some alleviation of symptoms and some prolongation of life, does not deal with any of the real, root causes of the diseases in question. Examples of such half-way technologies are cardiac transplantation, kidney and liver transplants, chronic renal dialysis, and, costliest and farthest of all from the underlying problem, the artificial heart. As long as the mechanisms of the diseases remain unknown, measures such as these, heroic in a sense for both the doctor and the patient, are all there is to offer. The existence of a workable artificial heart and the fact that it has already undergone one human trial, ought to be regarded as the most urgent of spurs for more basic research on myocardial disease. If we are compelled to develop this device for anything like the widespread use for which it will be demanded, the cost of health care will go through the roof.

The costs are going to go through the roof anyway between now and the end of the century, unless we can discover more effective and economic ways of coping with the problem of aging. Already, in 1983, 10 percent of the American population is over the age of sixty-five, and by the year 2000 that figure is expected to rise higher than 20 percent. If we are looking around for something to worry about for medical science and technology, we should place this problem high on the list of anxieties.

The array of specific questions to be asked is long and impressive; each question is a hard one, needing close and attentive scrutiny by the brightest and best of clinical and basic science investigators. As the answers come in—and sooner or later they will come in—there is no doubt that medicine should be able to devise new technologies for coping with the things that go wrong in the process of aging.

This is an optimistic appraisal, but not overly so, provided we go carefully with that phrase "things that go wrong." There is indeed an extensive pathology of aging, one thing after another gone wrong, failure after failure. The cumulative impact of these is what most people have in mind as the image of aging, and most people fear. But behind these, often obscured by the individual items of pathology, is a quite different phenomenon, normal aging, which is something else again, not at all a disease, a stage of living that can only be averted or bypassed in one totally unsatisfactory way, but is nevertheless regarded in our kind of society as a sort of slow death, everything in the world gone wrong.

The two aspects of the problem are quite different. One can be approached directly by the usual methods of science, but I am not sure about the other.

This list of pathologic events associated with aging is a long but finite one. Away at the top are the disorders of the brain leading to dementia, the single threat dreaded most by all aging people and by their families as well. Bone weakness and fractures, arthritis, incontinence, muscular wasting, cancer, parkinsonism, ischemic heart disease, pneumonia, and an increased vulnerability to infection, in general, all these and more. These represent the discrete, sharply identifiable disease states that are superimposed on the natural process of aging, each capable of turning a normal stage of life into chronic illness and incapacity or premature death. Medicine, and biomedical science, can get at them one by one, dealing with each by the established method of science—which is to say by relying upon the most detailed and highly reductionist techniques for research. If we succeed in learning enough of the still deeply obscure facts about Alzheimer's disease, we may have a chance to turn it around and, in the best of worlds, to prevent it. Lacking those facts, we will be stuck with no way at all to alleviate it, and no way to help.

So, reductionist science is the way to go, for as far ahead as we can see. If the science is successful, we can hope for a time when the whole burden of individual disease states is lifted away from the backs of old people, and they are left to face nothing but aging itself.

And what then? Will such an achievement by biomedical research remove aging from our agenda of social concerns? If old people do not become ill, with outright diseases, acute or chronic, right up to the hours of dying, are their health and social problems at an end? Should we then give up the profession of geriatrics and confine our scientific interest to gerontology?

Of course not. It is possible that there will be more things to worry about for old people than is the case today, and more old people coming to the doctor's office for help. Aging will still be aging, a strange process still posing problems for every human being faced with the prospect, and science will still have much to do and much to learn. But it will have to be a different kind of science, capable of comprehending more than the intricate mechanisms of individual diseases, large enough in its vision to perceive the existence of a whole person.

The classical reductionist approach will not do for this.

The word "holistic" was invented in the 1920s by General Jan Smuts to provide a shorthand for the almost self-evident truth that any living organism—and perhaps any collection of organisms—is something more than the sum of all its parts.

I wish "holism" could remain a respectable term for scientific usage, but, alas, it has fallen in bad company. Science itself is really a holistic enterprise, and no other word would serve as well to describe it. Years ago, the mathematician Poincare wrote that "science is built up with facts, as a house is with stones. But a collection of facts is no more a science that a heap of stones is a house." In any case the word is becoming trendy, a buzz-word, now almost lost to science. What is called "holistic thought" these days in medicine is similar to the transition from a mind like a steel trap to a mind like steel wool. "Holistic medicine," if it is anything, is an effort to give science a heave-ho clear out of medicine, to forget all about the working parts of the body, and get along with any old wild guess about disease. We do need another word to distinguish a system from the components of a system, but I haven't been able to think of one.

I went through my dictionary and got sidetracked, as usually happens. Leafing through, I ran across geriatrics and gerontology and I went looking for other cognates from the Indo-european root *ger*. Zarathustra I found, and Zoroastrianism. Aha, I thought, here's a new line between the wisdom of the ages and the wisdom of the aged, but then, when I looked more closely, I learned that Zoroaster and Zarathustra really meant nothing more than the owner of old camels.

But it is a line of thought to think about. The mind of an old person ought to be more than just the assemblage of sequential experiences, lined up one by one, seriatim, in the lifetime of that mind. And indeed it sometimes is, and when we observe this phenomenon—as we do in a few people in their eighties and nineties—we call it wisdom. The Indo-europeans had that word better set on its course: wisdom came from the ancient root *weid*, meaning simply "to see," and from that same root we have, as cousins, "wise" and "wit" in both its meaning, and "vision" and "advice," the things we should be expecting from old people, and also *Rig Veda*, a hymn to wisdom.

The trouble with looking at aging from the point of view of a scientist is that we are only accustomed to the reductionist way of looking. We can construct hypotheses about the possible mechanisms of senile dementia, and then set about looking for selective enzyme deficiencies or scrapie-like slow viruses in the brain, and that is surely the way to go. We can make up stories about the failure of cell-to-cell signals in the cellular immune system of aging animals, and examine at close quarters the vigor of lymphocytes in all their various stripes, probably at the end not only finding out what goes wrong in an aging immune system but also how to replace what's missing, and that is surely also the way to go. We have somewhere within grasp the information needed for explaining bone demineralization and fixing it. Given some luck and a better knowledge of immunology, microbiology, and the inflammatory process, we ought to be able to solve the problems of rheumatoid arthritis and osteoarthritis once and for all. We are learning some things about nutrition and longevity that we used to be ignorant about. If we keep at it, sticking to the facts at our reductionist best, we should be able to move geriatrics and medicine itself onto a new plane in science.

But we will still have people who must grow old before dying, and medicine will have to learn more than medicine knows today about what growing old is like.

The behavioral scientists, the psychiatrists and psychologists, the sociologists and anthropologists, and even the economists and the historians will all have a part in the work that needs doing, looking for data, piecing facts together, trying to make sense of the whole out of their separate parts of the problem. But their efforts, although probably indispensable, will not be enough.

I suggest that we need a reading list for all young physicians and investigators to consult at the outset of planning for a career in science, and particularly in the scientific study of aging. Young people cannot begin to construct hypotheses without having a ghost of an idea what it means to be old. To get a glimpse of this matter, you have to leave science behind for a while, and consult literature. Not "the" literature, as we call our compendium of research, just plain old literature.

One solid source book was written by the novelist Wallace Stegner, as good a writer as anybody around, better than most, who in 1976 wrote a book called *The Spectator Bird*, about a literary man and his wife getting on into their late sixties and early seventies. To qualify for my list you have to be old enough to know what you're writing about, and Stegner was the right age for his book. The novel is, or it ought to be, required reading for any young doctor. Indeed, *The Spectator Bird* is good enough to educate doctors for any specialty career, even cardiac surgery (come to think of it, maybe especially cardiac surgery).

Stegner was a friend of Bruce Bliven, the former editor of *The New Republic*, and Bliven is brought into the novel for a brief episode and a wonderful quotation. Someone asked him, when he had reached the age of eighty-two, what it was like to be an old man. Bliven said, "I don't feel like an old man, I feel like a young man with something the matter with him."

Another on my list is Malcolm Cowley and his book of personal essays called *The View from 80*. This is better than anything to be found in any textbook on medicine, infinitely more informative than most monographs and journals on geriatrics. Cowley writes with all the authority of a man who has reached eighty on the run and is just getting his second wind. He, like Wallace Stegner, was also attached to Bruce Bliven. He quotes something that Bliven wrote to friends three years later when he had reached eighty-five: Bliven wrote, "We live by the rules of the elderly. If the toothbrush is wet you have cleaned your teeth. If the bedside radio is warm in the morning you left it on all night. If you are wearing one brown and one black shoe, quite possibly there is a similar pair in the closet. . . . I stagger when I walk, and small boys follow me making bets on which way I'll go next. This upsets me; children shouldn't gamble."

Malcolm Cowley writes in great good humor, and most of the people he admires who have made it into their 80s and 90s seem to share this gift, but it is not always the light humor that it seems. An octogenarian friend of his, a distinguished lawyer, said in a dinner speech, "They tell you that you'll lose your mind when you grow older. What they don't tell you is that you won't miss it very much."

A few old people have written seriously about their condition with insight and wisdom. Florida Scott-Maxwell, once a successful actress, a scholar, and always a writer, wrote, "Age puzzles me. I thought it was a quiet time. My seventies were interesting and fairly serene, but my eighties are passionate. I grow more intense with age. To my own surprise I burst out with hot conviction. I must calm down. I am too frail to indulge in moral fervor."

Living alone in a London flat after the departure of her grandchildren for Australia, nearing her nineties, she wrote, "We who are old know that age is more than a disability. It is an intense and varied experience almost beyond our capacity at times, but something to be carried high. If it is a long defeat it is also a victory, meaningful for the initiates of time, if not for those who have come less far. . . ." When a new disability arrives, she writes, "I look about me to see if death has come, and I call quietly,

'Death, is that you? Are you there?' And so far the disability has answered, 'Don't be silly, it's me.' "

It is possible to say all sorts of good things about aging, if you are talking about aging free of meddling diseases. It is, I suppose, an absolutely unique stage of human life, the only stage in which one has both the freedom and the world's blessing to look back and contemplate what has happened to a life, instead of pressing forward for new high deeds in Hungary. It is also, obviously, a period for others, younger and filled with questions, to draw upon. It is one of the three manifestations of human life responsible for passing along the culture from one generation to the next. The other two are, of course, the children, who make the language and pass it along, and the mothers, who see to it that whatever love there is in a society moves into the next generation. The aged hand along the experience and the widsom, if they are listened to, and this, in the past, has always been a central fixture in the body of a culture.

We do not use this resource well in today's society. We tend always to think of aging as a disability in itself, a sort of long illness without any taxonomic name, a disfigurement of both human form and spirit. Aging is "natural," we say, just as death is natural, but we pay our respects to the one no more cheerfully than to the other. If science could only figure out a way to avoid aging altogether, zipping us straight from the tennis court to the deathbed at the age of, say, one hundred, we would probably vote for that. And, even if it could be accomplished by science—which is well beyond my imagining for any future time—society as a whole would take a loss. In my view, human civilization could not exist without an aging generation for its tranquility, and every individual could be deprived of an experience not to be missed in a well-run world.

It may be a mistake to use a word like "natural" for human aging. Senescence is not at all universal in nature, not even common. Most creatures out in the wild die off, or are killed off, at the first loss of physical or mental power, just as our own tennis stars begin to drop off in their late twenties, and virtually all of our athletes become old men in their professions, ready to be slaughtered before middle age. Aging, real aging, the continuation of living throughout the whole long period of senescence, is a human invention and perhaps a relatively recent invention at that. Our remote ancestors probably looked after their aging relatives much after the fashion of more recent aboriginal cultures, by one or another form of euthanasia. It took us a long time, and a reasonable, workable economy, to recognize that healthy, intelligent old human beings are an indispensable asset for the evolution of human culture. It was a good idea and we should keep hold of it, but if it is to retain its earlier meaning we will have to find better ways to make use of the older minds among us. The notion is abroad in the land that life, real life, stops at some age between sixty-five and seventy, and at that time society's main task is to find a place to store the people who have reached that age—off somewhere, somewhere pleasant if possible, but off for sure and out of sight.

We need reminders that some exceedingly useful pieces of work have been done in the past, and are still being done, by some extremely old people, in good health or in bad. Johann Sebastian Bach was a relatively old party for the eighteenth century when he died at age sixty-five, but he had just discovered a strange new kind of music and was working to finish *The Art of the Fugue*, an astonishing piece based on the old rules but turning the form into the purest of absolutely pure music. Montaigne was even younger at his death, fifty-nine, but he was an old man for the 1500s; he was still revising his essays and adding notes for a new edition, for which he chose the appropriate epigraph: "He picks up strength as he goes." In our own time, Santayana, John Dewey, Bertrand Russell, George Bernard Shaw, W. B. Yeats, Robert Frost, E. M. Forster (and I could easily lengthen the list by the score) were busy thinking and writing their way into their seventies, eighties, and beyond. The great French poet Paul Claudel wrote, on his birthday, "Eighty years old! No eyes left, no ears, no teeth, no legs, no wind! And when all is said and done, how astonishingly well one does without them!"

Of all the things that can go wrong in aging, the loss of the mind is far and away the worst, and most feared. Florida Scott-Maxwell wrote down what most old people fear the most and have in the back of their minds: "Please God I die before I lose my independence."

In spite of today's ignorance about so many different diseases, including those of aging, there is the most surprising optimism, amounting to something like exhilaration, within the community of basic biomedical researchers. There has never been a time like the present. Most of them, the young ones especially, have only remotely on their minds the connection of their work to human disease problems. They have an awareness that something practical and useful may come from their research, with luck, sooner or later, but this is not what drives the work along. What they are up to, and now bcoming supremely confident about, is finding out how things work. This is true for the immunologists, threading their way through the unimaginably complex network of cells and intercellular messages comprising the human immune system. It is true for the experimental pathologists and their biochemist colleagues at work on the components of the inflammatory reaction, with new controlling and destabilizing reactants turning up each month. The cancer biologists are totally confident that they are getting close to the molecular intimacies of cellular transformation, and the virologists among them are also riding high. Out in the front lines are the molecular biologists and geneticists, in possession of research techniques that permit them to ask, and then answer, almost any question that pops into their minds.

Perhaps these people will turn up items of information needed for the prevention of arterial disease and intravascular coagulation, or for the control and reversal of the immunologic reactants in charge of auto-immune disease. I have no doubt in my mind that the lines of research now open for cancer cell research will soon have laid out the names and numbers of all the players, and we will have choices to make among totally new kinds of pharmacology and perhaps immunology. I see no reason to think that neurobiology is in any sense the impenetrable science that it seemed to everyone years back, and the possibility of reaching a deep biochemical comprehension of the events gone wrong in schizophrenia and the manic-depressive psychoses is becoming a quite realistic prospect. The senile dementias, Alzheimer's and the rest, are beginning to look like proper biological problems, ultimately to be solved if the work is done by good people at a deep level.

This is one way to put the case. If you are feeling as I do, optimistic and hopeful but most of all fascinated by today's science, you can predict major advances in medical care, somewhere ahead in real time. But there is, of course, another way of viewing the scene. All the excitement, each confrontation of total surprise in the research, signifies that we really didn't know very much to begin with and perhaps we are just beginning to penetrate surfaces, with miles and miles to go and with decades of research still to be accomplished before any goal can be reached. It may turn out a much harder and longer task than any of us can guess, and the extent of our ignorance may prove much more profound than we know.

Either way you look at it, we are not about to change

the world, nor are we in any position to begin talking or worrying about changing a human being, much less the species.

Yet this is precisely what some people are worrying about, and for the life of me I cannot understand why. If we are extremely lucky in the research and can learn much more than we know today about the insertion of normal genes and the switching off of flawed genes in mammalian cells, we might discover how to correct a few heritable diseases in which single gene defects are the cause of disease. This could mean a technology for installing cells able to manufacture normal hemoglobin instead of sickle-cell hemoglobin, as well as methods for correcting a fair number of congenital enzyme deficienceis now responsible for death in infancy and early childhood. Inserting isolated genes into the somatic cells of a human being is a genuine possibilty for the future, although it is by no means near at hand. Changing the genes in germ plasm cells is quite another matter. I doubt that any accredited cell biologist has a good idea how to go about this formidable step, in experimental animals or in man. So far, and for a long time to come, the human genome seems to me quite safe from meddling. Nor is this likely to become a matter of either scientific interest or concern, if medicine can achieve success changing the code within the single genes of somatic cells. That step, if it works, would serve the whole therapeutic purpose in sickle cell disease and others of a similar nature, and I can see no harm or imaginable risk in trying.

But I think this is not really what is on the minds of people who are apprehensive about the manipulation of human genes—or what it has become fashionable to call (and I wish the term had never been invented) genetic engineering. I think they are worried about the prospect of altering human behavior or breeding up races of humans with different sorts of intellect. I doubt that anyone would have misgivings about the substitution of normal genes for the defective ones in congenital disease of children, if it were not for the apprehension that one thing will lead to another and before we are aware of it, the door will open to experimental modification of the mind of the human race.

To evaluate this risk, it is necessary to appraise the present state of our understanding of behavioral genetics.

I cannot think of a field in biomedical science about which less is understood by the public at large and by the scientists who study behavior, than the genetics of behavior—not just human behavior but animal behavior in general. In the case of humans, the problem is so complex and so obscure as to be almost beyond stating. It may well be that such behavioral traits as general intelligence or artistic talent or aggressive behavior are heritable characteristics, but it may equally well be that they are not, and there the matter rests. As far as genetic manipulation goes, it makes no difference, for there is nothing to manipulate.

It is true that one can use the technique of recombinant DNA to introduce an isolated gene from the cell of one animal into the genome of the cell of another. The recipient cell will then begin producing the protein or peptide coded for by the transplanted gene, and so will all the progeny cells of the recipient, but, mind you, this is all done in cell cultures, not in animals. One can accomplish the same end result by infecting an individual cell with a virus that happens to be carrying along bits of genetic information from a previously infected cell. The great usefulness of this refined technology is that it permits the most detailed chemical scrutiny of individual genes as well as their gene products. The ability to clone genes in this way has led to an entirely new method for analyzing the central genetic abnormalities at work in cancer cells, for instance, and is presently illuminating the puzzling capacity of lymphocytes to recognize with high specificity a near-infinity of foreign antigens. Sooner or later we will perhaps begin to understand the strange business of embryologic development and differentiation, employing techniques derived from the kinds of research being done on recombinant DNA today.

None of this has anything at all to do with behavior, however, so far as is known. The genetic controls that govern the construction of an embryonic brain, and those that will later determine how an adult brain functions, are absolute mysteries. Whether they are in any way related to the molecular genetics of protein synthesis is unknown. If it turns out, ultimately, that there are indeed genes that act as determinants for various aspects of human behavior or intelligence (which I am temperamentally inclined to doubt, in the absence of evidence one way or the other), it is inconceivable that single genes could act as such determinants. If it is so, myriads of different genes lodged in different chromosomes would have to be involved, even for an act as simple as a smile.

We are, I think, centuries away from this kind of science. I acknowledge that statements like this have been made before by elderly prophets of science, sometimes followed soon thereafter by cascades of astonishing new information to transform the field, but in this case I do feel secure. The human mind, I feel safe in asserting, is still away over our heads.

And even if I believed otherwise, and felt that recombinant DNA technology might someday be put to work to change human beings, and also felt (as I surely would) that such a thing should be prevented by any means possible, I would not know what line of research to stop. If such a thing were to come about it could only happen as the result of the purest kind of basic science, and when it happened it would have to come as a total, overwhelming surprise. Nobody could possibly have foreseen, in the early 1970s, the emergence of recombinant DNA as a laboratory technique. It happened as the result of a series of unexpected, unanticipated surprises in different research laboratories studying the structure and function of DNA. Step by step, each one taken in the dark, there came into view a procedure so simple and at the same time so beautiful, that everyone realized that biology would never again be the same. It has turned out to be the most important set of scientific observations to be made in biomedical research since Darwin, in my view, and it will turn out, further on and almost incidentally, to be the surest way into the problem of cancer. My point is that if you still view the recombinant DNA technology as a hazard to life, and wish we had never developed it, and were to think your way back to the steps that a committee might have taken in the late 1960s in order to foreclose the possibility, there is in retrospect nothing that could have been done except to put a stop to cell biology and genetics research across the board, in short, to stop biological science. And if that hypothetical committee had acted in that way, we would have no knowledge of oncogenes or their gene products today and no attractive way into the depths of the cancer problem.

You either have science or you don't. Technology is another matter. If ever it happens that the kind of information is at hand to make it possible to transfer human intelligence genes into the human genome, the undertaking itself could never be launched without the expenditure of a substantial proportion of any nation's gross national product. For all of my instinctive misgivings about politicians, I would trust the prudence and the good sense, if not the economic reservations, of any government. I can think of a great many things to worry about for mankind's future, with nuclear warfare at the top of the list, but I simply do not have the time or inclination to worry about cloning brains or behavior.

Even less do I intend to worry about cloning a whole

human being. Theoretically, of course, it is a possibility. All somatic cells contain within the DNA of their nuclei all the genetic instructions needed for reconstructing the entire, precise, same organism. In a skin cell, or a brain cell, the genes for being anything but that specialized cell type have been shut off, but the silenced genes are still there. It is known that the nucleus of a tadpole cell can be transplanted into the enucleated egg of an adult frog, and the combination will then proceed to reinvent exactly the same tadpole. There is some evidence, still highly controversial, that something like this can be done in mice. To extrapolate from this to the assertion that that same thing can be done with a human cell, combined with a neutral human egg, is a piece of science fiction already written out by Rorvik, but still a very long shot in real life. The question about the procedure is not so much *whether* it can be done as why anyone would ever want to try. I can imagine doing it, just as I once was persuaded to imagine the landing on the moon, and I do suppose that if you took the entire resources of the National Institutes of Health, including the Clinical Center, and spent ten times the NIH budget plus a large piece of the Department of Defense budget for the next ten or twenty years, you might get it done at least once. Suppose, then, you had cloned the genome of an eminent scholar and diplomat of exceedingly high intelligence, what would the expensive product be? It would be a newborn baby, much like any other, with a remarkable resemblance to the baby pictures of the scholar-diplomat, but then what? Unless you had immediately at hand the diplomat's original parents, themselves at the same age they were at his birth, plus all his siblings and aunts and cousins, the clone would be raised and nurtured in a different environment, speaking a different accent, thinking about the world in different ways from the moment of opening his eyes. Very well, you could get round this problem by first cloning the parents and relatives and waiting for these clones to grow up before starting on the xeroxed diplomat. But that would only be the beginning of the matter: schoolmates, teachers, friends, the odd person across the aisle in the subway, the novelists, journalists and moviemakers of the diplomat's childhood years, all of these would have to be duplicated, along with all their relatives and friends, unless all our ideas about the influence of environment on early childhood development are wrong. Before you had finished laying out the protocol you would find yourself planning to clone the whole world, bringing the environment back to its present, highly unsatisfactory state. As for the plight of the original subject, the parent scholar-diplomat, try to imagine his fate, doomed to see the upbringing and disciplining of himself all over again through adolescence, juvenile delinquency, acne, high school mathematics, insecurity, vanity and all the rest. I cannot imagine it, nor do I believe Congress would ever vote for it.

I am, of course, professionally biased, and I tend to overreact. I confess to a certain crankiness in my reaction to the warnings of imminent danger to society from basic biomedical science. Perhaps there are some risks ahead, even if my bias blinds me to them. But I will stay unrepentant, cranky, unwilling to listen, for as long as these small, improbable risks continue to engage the obsessive attention of a society that ought to be spending all of its worrying time, and all of its insomnia, on the immediate, close-at-hand danger of thermonuclear warfare. I will gladly join in the anxiety about changing the human species by science, once I become convinced that another kind of science is not about to wipe us out.

The Ghosts Are on the Wall: A Parable for Manipulating Life

KAREN LEBACQZ

In June 1983, a group of eminent clergy, joined by several scientists, signed a petition urging Congress to seek a ban on certain types of "genetic engineering."[1] It is not the first time in the last ten years that public concern over new technologies has run high, nor the first time a ban has been sought. Nonetheless, the clergy involved have been derided for naiveté and accused of "apocalypticism" and "panic."[2]

The current situation brings to mind a delightful song written by the Reverend Tom Hunter, a minister of the United Church of Christ.[3] The song is a child's lament, and the chorus goes as follows:

There's a monster in the closet
And the ghosts are on the wall.
So why do you keep on telling me
"There's nothing there at all"?
I know they're there; it's clear to me;
I see them every night.
And all I want is when I'm scared
You come and hold me tight.

As Hunter puts it, "Kids aren't the only people who know about them, nor the only people who want to be held tight." We all have monsters in the closet and ghosts on the wall—shadowy figures that threaten the security of our dreams. And, as any parent knows, saying "There's nothing there at all" is not a sufficient response.

New technologies for manipulating life include:

- artificial insemination
- *in vitro* fertilization and embryo transfer
- prenatal diagnosis (and selective abortion)
- genetic counseling and screening
- sex selection of offspring
- "genetic engineering" (gene splicing; recombinant DNA)

For some, these technologies offer hopes and promises of dreams fulfilled—the eradication of disease, the betterment of human life, the solution to nagging theoretical or practical problems. For others, they are more like ghosts on the wall: they threaten values and dreams held dear. Still others find in them potential monsters.

Is there a monster in the closet? Are the ghosts on the wall? Do we have reason to be scared? And if so, what would it mean to "hold each other tight" in the face of perceived threats to our dreams?

The Ghosts Are on the Wall

Each of the above technologies has generated unresolved legal questions, troubling ethical dilemmas, or persistent technical problems.[4] I call these the "ghosts on the wall." They cause uneasiness for many. Upon closer examination, however, some will prove to be merely the wind blowing the curtains or the streetlight casting shadows on the wall. It may take but a second glance to make them disappear. Others loom larger and linger on, requiring not just a second glance, but a third and fourth as well.

Among the more persistent are those related to gene splicing or "genetic engineering" (recombinant DNA techniques). Recombinant DNA techniques have been controversial since their inception. During the last decades, entire cities (e.g., Cambridge, Mass.) have been aroused to debate over the "ghosts" attached to this cluster of new technologies. While some today would consider such anxiety unnecessary, a number of "ghosts" remain.

Biohazards

The first ghost is the possibility of "biohazards"—the escape into the environment of organisms that might be harmful to the human community or the ecosystem. Indeed, at one point early in the development of these techniques, members of the scientific community called for a moratorium because of the possibility of such biohazards.[5] The establishment and acceptance of regulations and physical containment procedures have done much to dispel early concerns. However, the possibility of biohazards cannot be dismissed out of hand and remains a troubling ghost for many.

Human-Animal Hybrids

Sparked by science fiction scenarios, the thought of human-animal hybrids has raised many fears. And while the possibility is still considered fictional by many, the ethical concerns are genuine. Would such creatures be human, or animal? What rights would they have, if any? Are there some things we should not do even if we can? As the President's Commission for the Study of Ethical Problems in Medicine and Biomedical and Behavioral Research notes, the prospect of exploitive or insensitive treatment of new creatures cannot be dismissed as totally fanciful.[6]

Correction vs. "Improvement" of Humans

Current efforts are directed to correction of genetic defects. But is it possible to draw neat, tight lines between "correction" of defects and "improvement" of the species? Some argue that improvement remains fictional because traits such as intelligence are too difficult to control.[7] But others rightly note that the definition of "defect" is variable. Women were once considered "defective" by definition! What is now thought defective may some day be considered normal, and vice-versa.[8] Experience with eugenics programs earlier in this century should make us leery of attempts to define and eradicate "bad traits."

And who would define what is considered "defective"? Those who would receive the "treatments" are rarely those who define the "defects." Thus, in addition to problems of definition in what constitutes "correction" or "improvement," there are serious questions of justice and control.

Who Controls?

This brings us to the next "ghost." Genetic engineering techniques could place great power in the hands of the few. Public commissions and agencies may not be adequate to the task of "watchdogging." The wheels of bureaucracy and representation may grind too slowly to keep pace with new developments. The President's Commission, which is otherwise very circumspect in its conclusions, raised hard questions at this point and urged vigilance.[9]

Somatic vs. Germ Cell Research

Finally we come to the "ghost" that sparked the recent call for a ban: concern about research on the "germ cells" that control the transmission of traits through generations. While gene splicing on somatic cells may affect deeply—even irreparably—the life of an individual and her/his immediate family, splicing on germ cells (eggs and sperm) would affect not only that person but future generations as well. Ethical issues of justice between generations arise here. As one of the signatories of the recent petition put it, "Are we justified in taking risks with human life, especially when an entire lineage of descendants will be affected by our intervention?"[10]

The "Frankenstein Factor"

This cluster of ghosts has been summed up as the "Frankenstein Factor."[11] (The choice of term is a bit unfortunate, as it connotes a late-night movie instead of a genuine concern.) The Frankenstein Factor incorporates fears about changing the nature of the human race, creating physical hazards, unleashing a monster that will not be within human control, crossing natural barriers, and undermining the good of the whole human race in the interests of private profit or prestige. The speed with which events have unfolded and the unpredictability of direction and speed of new developments heighten these concerns. No matter how carefully research is done, there will remain—at least theoretically—the possibility of creating something that is both *other* than what was wanted and that represents a *threat* to human well-being on either a physical or a moral and social level. So long as there are unresolved technical, ethical, or legal issues, Frankenstein looms.

And so, some seek a ban. Others argue that no research would or should be done until safety can be well assured.[12] But recent experience with *in vitro* fertilization suggests that human research will be done before safety can be "fully" assured—indeed, even when many scientists still think additional nonhuman testing is needed. Two attempts at gene therapy in humans have already been made.[13] This supports those who argue that we will proceed before safety can really be assured. (Indeed, safety cannot be established until some research is done!)

Nonetheless, while the "ghosts" that make up the Frankenstein Factor are frightening to many, the issues they represent are not altogether new. What risks are acceptable in research; how important genetic lineage is and whether the germ line is inviolable; what justice between generations requires—these are not at root "new" issues, though they may be posed with new urgency and poignancy by current technological developments. Many if not most of our "ghosts" are shadows cast by familiar things seen in unfamiliar light. The fact that the light is unfamiliar should not blind us to the familiarity of the basic questions.

These questions are nonetheless real. They cannot be dismissed out of hand—"there's nothing there at all." New technologies do raise troubling and unresolved technical, legal, and ethical dilemmas. Some of these may prove to be only shadows on the wall, but some are genuine ghosts. Many of these issues have not yet been resolved and will not be for some time to come. Others may be irresolvable in principle. They should be taken seriously. They call for tough solutions and for considerably more ethical discussion than they have as yet received. The ghosts *are* on the wall!

The Monster in the Closet

Yet I do not think the Frankenstein Factor is our monster. Is there something else lurking in the darkness, waiting to devour our hopes and shatter our dreams? Is there a monster in the closet?

"Playing God"

In the face of the unknown—and perhaps the unknowable—past solutions do not always seem adequate. We have principles for doing research on human subjects, but are those principles adequate to deal with the new dangers and complexities of recombinant DNA research? We have established guidelines for medical interventions, but are those guidelines sufficient for "gene splicing" as a medical technique? Many people sense that with "genetic engineering" we are facing something that is difficult to describe but somehow *qualitatively new*. At such a point, the received tradition, no matter how valuable in other places, does not always suffice.[14] The failure of old models and paradigms—or the sense that they may not be adequate to the task—does appear to be a "monster."

The sense that we are faced with something qualitatively new is often expressed by the phrase "playing God." To "play God" is to extend human knowledge and control into hitherto uncharted territory. It is to know the (previously) unknowable, and with it come some concerns.

As described by the President's Commission, these concerns express on the one hand a feeling—akin to the Frankenstein Factor—that we will not be able to control the consequences of great powers of knowledge. On the other hand, they reflect a sense that humans are limited creatures who transgress some important boundaries if we try to know too much.[15] Perhaps the territory is not merely uncharted but inaccessible. Thus, either the *consequences* or the *nature* of the knowledge involved is taken to generate a prohibition. The concerns center in an apparent conviction that there are some things that we ought not to try to know (and perhaps that we will be punished if we do).

While most religious faiths encourage the growth of knowledge and the exploration of nature, most also see limits to human knowledge and control. Mythologies express these limits by describing humans as creature rather than creator. Now new technologies appear to thrust us into the role of creator rather than creature. While some would point out that we have always exercised control over the human future, including at least some measure of influence on the gene pool,[16] such assurances do not provide much comfort in the face of what is perceived as a qualitatively new phenomenon. It is this qualitative difference that invites the application of the label, "playing God."

Hence, debate rages over whether our explorations constitute "playing God" and whether "playing God" is wrong. Some decry the human tendency to play God. They fear harmful consequences and doubt whether we are capable of exercising such power wisely. Others claim that we *must* "play God"—we have no choice but to become the designers of our own destinies. Indeed, some even claim that this is our greatness. One way or another, they all assume that the monster—if there is one—lies in some inherent "wrongness" of trying to know certain things. To those who think that certain things ought not to be known, the "monster" lies in trying to know them. To those who think there are no limits to human knowledge, there simply is no "monster," and, therefore, no reason not to proceed with research.

Playing and Reality: A Lesson from Job

Does "playing God," trying to know, and exercising increasing control constitute our "monster"? I think not. From my perspective as a theological ethicist, the issue is wrongly conceived when it is posed in terms of "playing God." In Judeo-Christian tradition, neither knowledge nor control are wrong per se. Biblical figures are *bold* about knowledge, and the literature supports the goodness of knowing.[17] Humans are also charged with dominion or control over the environment.[18] Hence, the problem is not "playing God" if by that phrase we mean increasing knowledge and control.[19]

Yet, there is also need for caution. Nowhere is this better illustrated than in the Book of Job. Job is usually taken to be a story about human suffering and the problem of theodicy: "Why do the righteous suffer?" And so it is. But the story speaks not merely to the substantive issue of suffering, but to a fundamental methodological issue: *How adequate are human paradigms?* Job invites us to challenge the very ways in which we think. It invites us to question the adequacy of human logic and assumptions. It invites us to learn to ask different questions.

Job is an innocent man: "Have you considered my servant Job? . . . He is blameless and upright," declares God. Satan retorts that Job is good because he has not suffered. Then follows a litany of sufferings inflicted upon Job—the loss of his family and possessions, a series of painful illnesses—in an effort to make him abandon God.

Faced with seemingly inexplicable suffering, Job asks, "Why?" The bulk of the story consists of arguments between Job and his (purported) friends regarding whether it is *right* that Job should suffer. Is his suffering merited? Job's friends argue that he must be guilty or he would not suffer.[20] Job argues that he is not guilty, declares his suffering unmerited, and accuses God of injustice.

On the question of Job's guilt, Job and his friends take opposite stands. Yet underneath their disagreement they share a common assumption. All think that only the guilty *should* suffer. Since Job in fact suffers, his friends jump to the conclusion that he must be guilty. Since Job knows that he is not, he declares his suffering unjustified. But all operate on the same paradigm: only the guilty may justifiably be made to suffer.

Ultimately, God vindicates Job's understanding of divine justice. But God also rebukes Job. And in the rebuke lies a lesson for us. To Job's demand for just treatment, God responds:

> Where were you when I laid the earth's foundation? . . . Have you ever given orders to the morning? . . . Have you journeyed to the spring of the sea? . . . Who cuts a channel for the torrents of rain? . . . From whose womb comes the ice?[21]

The litany goes on and on: "Where were you when? . . ." "Can you do this? . . . " "Who did that? . . ." These passages are among the most powerful and poetic in the Bible. And in the midst of the litany are these words: "Who has a claim against me that I must pay? Everything under heaven belongs to me" (41:11). Job has approached God with the language of *justice*. God responds with the language of *creation*. God's response changes the very rubrics of the conversation. As the story is told, God in essence declares, "I am beyond your categories." ("Who has a claim against me that I must pay?") The rebuke constitutes a challenge to the paradigm accepted in common by Job and his friends.[22] It suggests that Job, though righteous, has asked the wrong question! Within the limits of human vision, within the limits of the paradigm presented by Job and his friends, Job is right. But the paradigm itself is wrong. The human categories are too limited to contain all of reality.

The Monster in the Closet: Paradigms of Knowledge

And so it might also be with us. While correct within the limits of human vision, we might nonetheless be asking the wrong questions. Like Job, our very paradigms may be the problem.[23] Job was a righteous man, "blameless and upright." And yet, his vision was limited, his question wrong. I doubt that we can claim the same righteousness Job did. (Could it be said of us that we always gave to the poor, acted justly, and honored the socially outcast?!) If even the righteous can ask the wrong question, how much more so might we?

If there is a "monster in the closet," then, it may not be the new technologies themselves, nor even the powers they would give us. It may not be the Frankenstein Factor—the possibilities of uncontrollable results that might harm the human community. Nor may it be the tendency to "play God" as this is usually interpreted—the notion that there are forms of knowledge that humans should not try to possess and arenas into which humans should not extend their powers.[24]

The problem is not "playing God," but confusing play and reality. The problem occurs when we forget that we are only "playing" and that God remains God in spite of all our play. To "play God" is not to "be God."

Job and his friends appeared to be on opposite sides of the argument and yet shared a basic paradigm. This paradigm reflected the received tradition but was not adequate to the current issue. In like manner, many engaged in the current debate are on opposite sides when it comes to social policy recommendations as to whether research should continue. Yet all share a received tradition. The real question is whether this received tradition is adequate to the current task.

For instance, some urge caution on the grounds that we do not know enough to act wisely. Others urge action on the grounds that we do know enough. Still others urge action precisely so that we can come to know enough. While they differ on policy proposals—with some supporting a ban and others not—all share the same paradigm in which caution or action depends on knowledge. What we should *do* depends on what we *know*.

Furthermore, in our received tradition, "knowledge" is defined rather narrowly. Both in Western science and in Western ethics, reason, scientific evidence, and logic are largely what is meant by "knowledge." Thus, we may act or not depending on whether we have accumulated enough data of certain types and drawn sufficiently logical conclusions from that data.

This approach is exemplified by the generally excellent report of the President's Commission. Cautious and careful, attuned to subtle distinctions and innuendoes, modest and amazingly thorough for its length, this report is in some ways a model for ethical analysis. Yet throughout its discussion, one gets the uncomfortable feeling that too much depends on logic and rationality. The report calls for "dispassionate appraisal" and "sober recognition."[25] It looks for the "rational kernel" in fears and admonitions.[26] No room is made for passionate appraisal or for a form of knowledge that goes beyond rationality.

In its discussion of "playing God," for instance, the Commission sorts through the logical alternative meanings of the term, finally concluding that fears of "playing God" must be at root fears of the consequences of exercising great power.[27] The pattern of logical analysis and elimination of possibilities that is used to arrive at this conclusion suggests that the Commission operates on a particular paradigm. This paradigm assumes that all the alternatives can be located and each can be dismissed in turn. The dismissal requires logic and the rational use of evidence. There is nothing larger than or outside of the particular alternatives. Hence, once each alternative is demonstrated not to hold water, all objections disappear.

But is it possible that all of the logical alternatives could be demonstrated not to hold up to rational analysis and *yet* that there could be reason to worry? I think so, for I am not convinced that the process of logical elimination is the only proper tool with which to approach the issue.

Two problems are hidden by the Commission's trust in human logic and rationality. First, the report gives the impression that humans can set aside their biases and find an "objective" answer. It gives the impression that there is such a thing as unbiased thought. This ignores the deep biases that are built into Western forms of thought, which acknowledge only certain types of insight and inference as "knowledge."

And this points to the second problem. In trusting logic and rationality for answers, other modes of insight are left out of account. Prayer, meditation, dreams, the insight that comes as an intuitive response—all these and many more are recognized as souces of knowledge in other traditions but largely ignored in our own. To depend on the forms of knowledge honored in this culture is not only to ignore the biases built into those forms but is to ignore other sources of insight as well.

I think there *is* a monster in the closet. But the monster is not simply the possible deleterious consequences of new technologies. Nor is it merely the threats posed to deeply held values. It is our tendency to assume that even as we give lip service to human fallibility and ignorance,[28] we nonetheless can trust both in our limited forms of knowledge and in the paradigm that posits knowledge as the key to action. The monster is our tendency to give allegiance to human thought itself—to logic and rationality, to our ways of posing questions, to our presumptions about what it means to "know" something. The "monster" is our tendency to think that God—or reality—must be conformed to *our* paradigms and *our* ways of thinking.[29] In Biblical tradition, putting trust in anything other than God is a form of idolatry. This includes putting trust in human reason.

To say this is *not* to say that it is wrong to try to know. Nor am I suggesting that there are "sacrosanct" areas of life into which we should not intrude. Indeed, I think we should try to know everything we possibly can. But as we do so, we must pay careful attention to the adequacy of the paradigms we use.

Nor am I suggesting that we should give up reason, science or ethics! We still need reasoned argument. There are areas of great achievement in science and medicine, and these should be honored. The received tradition has value. But when we stand on the threshold of something that is *felt* to be qualitatively new, the received tradition may not be adequate and we may need to be open to new insight. We should not *assume* that we can trust our typical paradigms or received tradition to be adequate. In making that assumption, we may create a monster in the closet.

And All I Want Is, When I'm Scared
You Come and Hold Me Tight

If the monster lies in our limited thought, then how do we "come and hold each other tight"?

There are some ways that do not work! First among these is refusing to see the issues or belittling the concerns—"there's nothing there at all!" Those who deride the clergy for panic or assume that they were "taken in"[30] by Jeremy Rifkin resort to ad hominem arguments that ignore both the real issues and the substantial expertise of some of these clergy. The clergy's concerns are not recent, nor are they the simple result of one man's campaign.[31] Belittling the concerns is not the proper approach.

At the same time, a ban on research may also not be the best way to "hold each other tight." If what has been said above is correct, then the question is not so much what we *do* but how we *think* about it. The problem is not that we would proceed with research, but rather the paradigms on which that research depends. Still, the clergy have some force in calling for a ban—or at least a moratorium—since thinking new thoughts and being open to new modes of insight is not an easy task. Time will be needed.

Finally, I am skeptical about the establishment of another National Commission. Even a commission of diverse representation may remain limited in its ability to get perspective. Job and his friends were a diverse group. They differed greatly. And yet they failed to see that all assumed the same underlying paradigm. Even a diverse Commission may fail to see the basic assumptions out of which it operates. Like Job, it may be right and yet also wrong. It may ask the wrong questions.

A Hermeneutic of Suspicion

How, then, do we "hold each other tight"? First, we need a "hermeneutic of suspicion" that goes to radical depths. We should be suspicious not simply about what we might do with new technologies, nor about the consequences of their development and use, as those who signed the controversial petition appear to be. We should be suspicious not simply about who will control the new technologies, as the President's Commission urges. Rather, we should be suspicious about the very ways in which we *think* about the issues involved. We should be suspicious not only about what we should *do* with new knowledge, but even about what *constitutes* knowledge. Can we even ask the right questions? This is the point of the book of Job.

How do we get such a "hermeneutic of suspicion"? If one is a woman or a member of an ethnic minority in this country, it is not hard to come by suspicion about the ways in which we have been told the universe works—and what our place in it is! We have long been suspicious of the "logic" and "rationality" paraded before us by those in power in white, male-dominated institutions. Similarly, those from the so-called third world hold deep suspicions about dominant Western paradigms. One way to find a "hermeneutic of suspicion" is to take more seriously the voices of the oppressed both here and abroad.

This has been difficult for leaders in the West to do because those who cry out in pain or oppression rarely share the basic assumptions of the dominant class. Like the child's cry in the night, the very voices that we need to hear often operate from a stance beyond (though not excluding) rationality. They draw on sources of insight other than logical argument.[32] And like the parent responding to the child, it is tempting to say "there's nothing there at all" and to dismiss the concern because it seems not to fit our paradigm. How, then, can we learn to hear different rubrics and to be suspicious of our own?

A Parabolic Paradigm

At this point the Bible has much to offer us. Biblical authors knew how difficult it is for people to transcend their traditions and paradigms. At points where the received tradition was not adequate and new insight was needed, they often turned to parables. Both the Old and the New Testament are permeated with *mashal*—"parables" or "riddles."[33] A parable is "a picture puzzle which prompts the question, 'What's wrong with this picture?' "[34] A parable is a story—a picture. The listener is invited into the story and identifies with one or another of the characters. And then something happens—a twist of events, something unexpected or odd, a question posed. And the listener is forced to ask, "What's wrong with this picture?" In the asking, or answering, the listener's world-view is often turned upside down. Familiar paradigms are shattered.

Jesus told many parables. We have often wrongly taken these to be little stories that give us rules to live by. For instance, the parable of the "Good Samaritan" is taken as a lesson about helping our neighbor. But this would not have been the case for Jesus' audience. His audience was Jewish. When he begins, "A man was going down the road," they undoubtedly picture a Jewish man (perhaps even putting themselves into that man's place). They identify with the one who was robbed and hurt. The question they are asking, then, is "what will happen to me?" and "who is going to help me?" The twist in the story comes when they are told that a Samaritan stopped to help. Jews and Samaritans were mortal enemies. The parable suggests that the one who can be counted upon to help them is an "enemy," a Samaritan. This is like telling a Black person that the one who will help him or her is a member of the Ku Klux Klan—or telling Ronald Reagan that the one who will help him is a Communist! To accept this requires a radically different perspective and shatters the listener's assumptions.[35]

The Book of Job, taken in its entirety and including God's rebuke, is a parable. The rebuke is unexpected. Job is right—he *is* blameless and innocent. Even God has said so. And yet, God rebukes Job. The rebuke jolts the listener and forces a new perspective—in this case, the shift from the language of justice to the language of creation. The parable forces Job to look at the paradigm he brings and the questions he asks.

In short, parables bring us face to face with the limitations of our paradigms, our ways of thinking about the world. They provide a way to "hold each other tight" *not* by clinging to the rules and securities that we have but precisely by opening up new possibilities and new questions.

There's a monster in the closet
And the ghosts are on the wall.
So why do you keep on telling me
"There's nothing there at all"?
I know they're there; it's clear to me;
I see them every night.
And all I want is when I'm scared
You come and hold me tight.

New possibilities for manipulating life have put ghosts on the wall. Saying "there's nothing there at all" is not an adequate response. Lest we find—or create—a monster in the closet, let us "hold each other tight" by being suspicious of the limits of our thought and opening ourselves to the parables of life.

NOTES

1. *Time*, 20 June 1983, p. 67.
2. C.f. *Discover*, August 1983, p. 6; Bernard D. Davis, "Cells and Souls," *New York Times* 28 June 1983; and Horace Freeland Judson, "Thumbprints in Our Clay," *The New Republic*, 19 & 26 September 1983, pp. 12–17.
3. Tom Hunter, "Monster in the Closet," c. 1977. Used by permission of author. (Album: "Comin' Home," Long-Sleeve Records, Mill Valley, Ca.)
4. C.f. Karen Lebacqz, ed. *Genetics, Ethics and Parenthood* (New York: The Pilgrim Press, 1983).
5. President's Commission for the Study of Ethical Problems in Medicine and Biomedical and Behavioral Research, *Splicing Life: The Social and Ethical Issues of Genetic Engineering with Human Beings* (U.S. Govt Printing Office, November 1982), p. 11
6. Ibid., p. 58 f.
7. Judson, "Thumbprints in Our Clay," p. 17.
8. President's Commission, p. 22.
9. Ibid., p. 72.
10. J. Robert Nelson, "Genetic Science: A Menacing Marvel," *The Christian Century*, 6–13 July 1983, p. 637.
11. Originally coined by Willard Gaylin, President of the Hastings Center (Institute for Society, Ethics, and the Life Sciences), New York, the term is used by the President's Commission, p. 14 and passim.
12. C.f. President's Commission, p. 61; Bernard D. Davis, "The Two Faces of Genetic Engineering in Man," *Science* 219, no. 4591 (25 March 1983): p. 1381.
13. Judson, "Thumbprints in Our Clay," p. 17.
14. Walter Brueggemann, *The Creative Word; Canon as a Model for Biblical Education* (Philadelphia: Fortress Press, 1982), p. 43. "The Ghosts Are on the Wall" was originally drafted before I encountered this remarkable book; it seems fitting, however, to make reference to it in several places.

15. President's Commission, p. 54 f; Nelson, "Genetic Science," p. 636.
16. Davis, "Cells and Souls."
17. Brueggemann, *The Creative Word*, p. 81.
18. C.f. Gen. 1:28–29.
19. However, the phrase "playing God" is a bit unfortunate, as it suggests that the issues may be "playful."
20. One argues that the upright are never destroyed; hence, Job could not be upright or he would not have been destroyed (4:12f). Another claims that Job is being punished for lack of faith and inadequate worship (11:1f). A third argues that Job's very complaints of injustice are angering God and bringing on his punishment (15:25f).
21. Job 38 f, passim, New International Version.
22. I use the term "paradigm" here in a broad sense to indicate a pattern or archetype of thought. Thus, my usage is perhaps broader than the current usage, e.g., in physics, but is not disconsonant with it.
23. Note, for instance, that we have persisted in calling God "father" and using almost exclusively male referents for God in spite of the obvious reference in the passage cited to God's "womb"! This is an indication of how deeply ingrained paradigms can become, even in the face of considerable evidence to the contrary.
24. Indeed, I believe that the "monster in the closet" has been wrongly defined to be the attempt to control nature or extend human powers. In Jeremy Rifkin's controversial book *Algeny* (New York: The Viking Press, 1983), the tremendous powers of control brought about by genetic engineering are lifted up as dangerous and potentially wrong in themselves. While I agree with Rifkin's desire for a stance of "humility," I think he has established a false "either-or" in which humility means refusing control. "Either-or" thinking is itself one of the "paradigms" that accompanies scientific, logical thought; to establish an "either-or" between humility and control therefore seems to me to be creating the problem, not solving it.
25. President's Commission, pp. 58–59.
26. Ibid., p. 58.
27. Ibid., p. 58.
28. Ibid., p. 59.
29. In a day when we have become aware of the limits of some of our basic paradigms in physics, it is surprising that we continue to rest so assured about paradigms and perspectives in other fields.
30. *Discover*.
31. Many of these clergy were influential in getting the President's Commission established and in putting genetic engineering on its agenda.
32. Brueggemann argues that in the prophetic mode in the Bible, "new truth is likely to be a *cry* from 'below,' not a *certitude* from 'above.' New truth from God is likely to come as a cry and a protest of the weak, the powerless, the disinherited ones." *The Creative Word*, p. 63.

33. Robert H. Stein, *An Introduction to the Parables of Jesus* (Phliadelphia: Westminster Press, 1981).
34. R. W. Funk, *Language, Hermeneutic, and Word of God* (New York: Harper & Row, 1966), p. 158.
35. C.f. Brian O. McDermott, S.J., "Power and Parable in Jesus' Ministry," *Above Every Name: The Lordship of Christ and Social Systems*, ed. Thomas E. Clarke (New York: Paulist Press, 1980); John Dominic Crossan, "Parable, Allegory, and Paradox," *Semiology and Parables: An Exploration of the Possibilities Offered by Structuralism for Exegesis*, ed. Daniel Patte (Pittsburgh, Pa.: The Pickwick Press, 1976), pp. 247–82; and Pheme Perkins, *Hearing the Parables of Jesus* (New York: Paulist Press, 1981).

Bio-Engineering: Short-Term Optimism and Long-Term Risk

CHRISTIAN ANFINSEN

Whatever our background or occupation, most of us do not hesitate to make profound statements on philosophical and ethical matters. The subject of our current conference—"Manipulating Life; Medical Advances and Human Responsibility"—is ideal for crystal ball-gazing. The technology of gene transfer and modification is very much in a state of flux, and the medical and evolutionary consequences are unpredictable. It would be difficult to rule out any reasonable prognostication.

Quite by accident, I came across a lecture by an old friend, Ernst Chain, who shared the Nobel Prize for the discovery and production of penicillin, in which he gives his opinion on the use of genetic engineering in man. The lecture was given just six years ago, in 1977. I would like to quote from the remarks by this eminently reasonable man:

> There exists no method, at present, nor is there the likelihood that one will be discovered in the foreseeable future, by which it would be possible to alter the nucleotide sequence, and thereby the genetic properties, in any gene of any mammalian cell in a *controlled* manner which could be termed "genetic en-

gineering." Any speculations that such a process may be near at hand and could influence the heredity of man must be dismissed as science fiction.

The unpredictability of discovery in a fast-moving field such as mammalian genetics is underlined by recent observations in exactly the field of research that Professor Chain considers in this quotation. Most of us have read or heard about the production of "giant mice" by genetic engineering. In these experiments, the gene for mouse growth hormone was injected into mouse eggs that had been fertilized, and the eggs were then allowed to mature *in utero*. The resulting animals apparently looked more like rats than mice, and the whole experiment is not only interesting but somewhat amusing. This experiment emphasizes how quickly the development of new techniques for the preparation and study of biological materials can lead to astounding biological results. The next step, of course, will be a set of similar experiments on cattle and sheep, and other animals of interest to man. The very existence of the "giant mouse" experiment implies that similar experiments could be done with humans. The controlled injection of the gene for human growth hormone into fertilized human ova could easily lead, in twenty years, to a Minnesota football team made up entirely of nine-foot players. Although the experiments I have just mentioned have a certain humorous quality, I will, in this lecture, take a moderately gloomy view of the possible undesirable consequences of genetic engineering as applied to man.

Being a scientist myself, I have some insight into what makes a research worker tick. The qualities that characterize and motivate a good scientist do not necessarily have any bearing on the ethical or sociological sequelae of discovery. An investigator who is worth his salt will attack an interesting problems for its own sake, and with increasing enthusiasm as the project proceeds. I am sure, for example, that Enrico Fermi, when he discovered the consequences of slow neutron bombardment of atoms—or Otto Hahn, when he studied the fission of heavy atoms—did not for a moment think ahead to the Hiroshima bomb, or for that matter, to the difficulties that sooner or later might present themselves in connection with nuclear power. In my view, as long as there are medical advances to be made, and diseases to cure or alleviate, research workers in the biomedical field will continue to explore the promising leads.

Before considering the many beneficial aspects of genetic engineering, I would like to continue just a little longer on the topic of the responsibility of the scientist for his own discoveries. I am sure that any reasonable scientist would fight against the use of his or her research products for thoughtless or criminal ends. Although no set of rules has yet been devised for the control of experimental genetic manipulation of human beings, I would guess that such a code will eventually be devised, and will, of course, depend on the cooperation of investigators in the field. The unfortunate fact is that the public cannot imagine the rate of progress in a field such as genetic engineering, and it might be quite difficult to devise supervisory rules and regulations that would be sufficiently flexible and *au courant* to keep up with advances in knowledge and clinical trial.

Even with a high level of understanding and cooperation, one can imagine situations in which control would be deliberately sabotaged. Consider, for example, the deranged scenarios in the concentration camps in World War II. If current techniques had been easily available, experimenters involved with captives would almost certainly have tried the modification of human fertilized eggs by exposure to purified genes or to heterogeneous mixtures of DNA fragments.

This technique which, at the present time, is somewhat

limited by the shortage of normally shed ova, could easily be speeded up by the use of ova obtained from ovaries following surgical procedures. I apologize for such a gory example, but I do feel that rules will have to be promulgated, and some worldwide control system will have to be devised to restrict misuse. At the moment, world peace and the Golden Rule are about all we can suggest. I must add that, even though I feel that regulations must be invented, I do not predict more than mild success. The situation is very much a case of Murphy's Law, which states that if something bad can happen, given enough time, it will.

For a change in pace, let me review some of the good things that have happened as the result of the development of the very sophisticated procedures of gene cloning in bacterial cells. The procedures for the controlled introduction of specific human genes in rapidly growing species such as *Eschericia coli*, which can subsequently manufacture and secrete the material of interest, are now the basis of a flourishing industry involved in the production of various peptides and proteins that can serve as weapons against human disease with essentially no risk.

One very popular item of interest, at present, is the protein known as interferon, which is produced by a number of different types of cells in the human body when these cells have been exposed to viruses. My colleagues and I spent about ten years working on the isolation and purification of interferon from human cells in culture, but the introduction of the cloning methods more or less outmoded this more classical approach. When the purified gene for interferon is introduced into the genome of the bacterial cell, along with the proper genetic start signals, the cells secrete fairly large amounts of the substance, and purified interferon can be obtained by relatively simple procedures.

Interferon promises to be of use in the control of a number of viral diseases, including hepatitis, conjunctivitis, and possibly some forms of cancer. The availability of interferon also makes possible the large-scale production of antibodies against this substance, which can be used to remove interferon from the circulating blood in certain diseases such as lupus and rheumatoid arthritis where the levels are abnormally high, and may be involved, somehow, in the dangerous aspects of these diseases.

Another substance that has been prepared by gene cloning in bacterial cells is insulin. The cloned gene is for human insulin, since one of the problems that can arise in diabetes is the dangerous immunological effect of insulin taken from the pancreas glands of other species.

Most of the materials so far produced in bacterial culture have been fairly small proteins of polypeptides. It seems certain that the procedures will be improved to a point where large human proteins can be made in quantity. Many human diseases result from the absence of a particular gene, and therapy for such gene-deficiency diseases might well turn out to be the introduction of human proteins made by the cloning method.

In some cases, genetic diseases attributable to the absence of critical genes in the chromosomes of patients involve only one, or a very few tissues. Thus, for example, sickle cell hemoglobin is produced by abnormal cells in the bone marrow, lacking the normal hemoglobin gene. These cells produce a form of hemoglobin that causes severe damage, pain, and, frequently, early death because of its propensity for precipitating out of solution within the red cells of the patient at low oxygen pressures. A large number of single-gene-deficiency diseases are known that result from the inability to produce a specific protein product in the cells of one tissue such as brain or pancreas. In such cases, one would like to try to introduce the correct gene into the cells in question without having this genetic material inserted into other cells of the body that are not associated with this particular gene-protein system. Research is underway on techniques that might permit this targeted delivery of genetic material. The experiments are based on the fact that all cells of the body have, on their surfaces, specific receptor sites, which can be recognized in a unique way by the proper molecule. One could, in principle, attach the missing gene to a substance that would recognize the surface of the cell type in question, and thus deliver the gene to its proper location. This type of manipulation carries with it the possibilities of bad results, as well as good. Should such a circulating gene be taken up by tissues other than the one for which the targeted gene was devised, it could create a genetic defect worse than the defect the gene therapy was intended to correct.

Lewis Thomas, in his book *The Medusa and the Snail*, has pointed out that current advances in the understanding of biological phenomena make it possible to begin thinking about a human society free of disease. Because of the nature of science and scientists, research directed at the achievement of such a medically utopian state will undoubtedly continue in an inexorable (and a scientifically fascinating) way, not only through the use of the newer methodologies for manipulating and modulating life, but also through the more classical techniques of synthesis and testing of drugs. The concept of a world nearly free of disease, however, considered in the context of our current sociological inadequacies, is frightening. The uninhibited increase in the population of the world, and the almost inevitable increase in hunger and crowding, are the real problems. It is unfortunate, therefore, that the study of the modulation of human behavior and of the rate of human multiplication are not subjects with the level of popularity enjoyed by biomedical science and the curing of the sick.

The field of biotechnology is most heavily concerned with the bio-engineering of bacteria to produce useful items such as interferon. Surprisingly little effort or money is spent on using the new biotechnology in areas of food production and population control. We tend to forget that the number of sick people in the world is really rather small when compared with the number of relatively healthy people who go to bed hungry every night. For this reason, I strongly advocate a greater attention to the application of biotechnology to our food sources. The subject does not have the "crowd appeal" of cancer cures or even "giant mice," but perhaps the manipulation of plant rather than human life might ultimately prove to be the most significant direction for research to take.

Let me return for a moment to my central theme: the misuse—or overuse—of scientific discovery. Scientists are motivated, generally, by (1) intellectual interest, (2) the thought of making a living in a pleasant profession, (3) the desire for satisfaction of personal ambitions and the need for accolades, and (4) societal forces most frequently stemming from military and economic pressures. I maintain that we will find it impossible to legislate against scientific research of any particular sort, including genetic engineering, by simply admonishing or lecturing to people who have no real desire to listen. It appears to me that only political, economic, and occasionally ethnic or religious pressures can force large-scale changes in behavior patterns. It is possible that the physical, mental, and psychological makeup of the human animal is such that there will be no solution to the dilemma of how to preserve the human species in the backwash of his own inventions.

One factor that might help is the control of population and, if possible, diminution to a much smaller level than now exists on the surface of the planet. Since it would be preferable to achieve a greatly diminished population by some means other than systematic nuclear bomb-dropping, an international effort to educate the people of the world in the direction of a negative population rate should be a first priority. Parenthetically, there might be considerable resistance to such a move by those whose fortunes depend on an ever greater population of purchasers.

I feel free, in the context of the conference title, to speak of population control, because much of the problem might be solved by products of contemporary biomedical research and biotechnology. There is, of course, the eternal problem of people in the impoverished agricultural areas desiring to have large families, particularly strong males with a liking for digging in the fields. We might need entirely new answers to world food production and distribution. In this connection, I am reminded of the conversation I had recently with Professor Dennis Powers of The Johns Hopkins University, who is intrigued with the use of the growth hormone gene in the production of "giant fish." He points out that numerous countries, including China, Israel, and others raise a great deal of animal protein—in the form of fish—in ponds. A beautiful blend of nutritional need and modern bio-engineering would be the introduction of the gene for fish growth hormone into the fertilized eggs of the edible species, with the subsequent establishment of a biologically competitive subspecies. In modern biology almost anything goes.

Some months ago, Professor Esbjornson, our organizer, wrote to each member of the panel, listing some preliminary reflections on the theme of the conference. Many of his reflections took the form of questions I should like to try to answer along the lines of the remarks I have just made.

Professor Esbjornson asks, "What do we want to know about life processes?" "Everything," I would say. He asks, "Are there limits we cannot or should not violate?" I would answer this question "no" in the context of basic research aimed at understanding more about our universe and about the nature of living things. At that time when the *application* of certain classes of scientific observations is contemplated, careful control by suitable panels of academic, legal and governmental experts should certainly be imposed. "Are we disturbing the universe by our probes?" I would say that we certainly are. Let me once again mention giant mice and nuclear bombs. "Are we capable of carrying the burden of responsibilities for the new knowledge we are gaining?" I would answer "probably not" to this question. In the short run, and during times when our planet is not being wracked by world wars or devastating epidemics, we seem to be able to manage.

I sincerely believe that we can maintain adequate surveillance of the application of bio-engineering to human beings so long as the human hunger for power and material gain does not become overwhelming. I think it would be difficult and inadvisable to attempt to control the normal progress of scientific research and application. In my view, the problems that arise because of advances in human knowledge in general, and biotechnology in particular, must be dealt with by those among us who are properly experienced in the moral and legal arenas of our society.

What's So Special about Being Human?

WILLARD GAYLIN

My father, as a diligent parent, had taught me that mature people must take responsibility for their failures. He neglected to tell me that mature people must also take responsibility for their successes. In medicine today, we suffer from our successes. And physicians are confused. Like martyred mothers, they ask: "After all we've done for you, how can you treat us this way?"

My answer is that it is precisely because of our current successes that we are "treated this way." As Lewis Thomas pointed out, until very recent times, medicine was essentially a care and comfort profession; it was not a lifesaving profession. As a reward for our achievements, we are regulated, criticized, overseen, legislated, and litigated to a degree that had never been anticipated. And rightly so. For with success medicine entered the realm of ethics and esthetics. In these areas physicians hold no special authority. Their decisions are as subjective and value-laden as those of their patients. They must cede, therefore, some of the decision-making to the patient or the patient's surrogates—the legislatures and the courts. In our impotence, we were protected against philosophical and theological considerations. With success came choice: with choice, dilemma. Two examples will suffice.

Any head trauma sufficiently extreme to destroy all our aspects of humanity—the capacity to perceive, feel, and respond—would normally also take the vegetative processes along with it. You would cease to breathe and you would die. With the development of the respirator we cleaved the natural and merciful bond between the vegetative and the human functions. Now we were forced to address the questions of whether that which we were keeping alive on a machine was still human (even the way you frame the question indicates your bias) or whether that human being that we were sustaining on the machine was still alive. We were forced to explore questions as fundamental as the definition of life and living, and this eventually led to a need for a legal definition of death. The nature—the quality—of living and life emerged as a crucial issue for consideration by physicians ill-prepared to deal with such questions.

Similarly, at one time surgery was quite limited. With antisepsis we began to evolve the kind of elaborate surgery we could never have anticipated before. At one time, survival days, a quantitative phenomenon, was the index of proper treatment for a disease. A hundred survival days for surgery was preferable to fifty survival days for medication. But when surgery became so extensive that you could remove an entire face, surgeons began to question whether those hundred days were indeed as valuable as fifty days which allowed re-entry into the community and the maintenance of social and business life. So surgeons and the rest of us were forced to talk about not just the quantity, the amount of life left for us, but the quality of life that remained. These were the introductory problems that dominated discussion in the earliest days of bioethics.

With still newer developments in the exploding capacities of the biological revolution another set of concerns emerged. The second group of problems focuses on the nature of our nature. When we began to be able to influence human behavior via the use of drugs, electrodes, and direct surgical intervention in the brain—a panic occurred. There was enormous suspicion about our capacity to change behavior through such manipulative devices and invasive mechanisms. After three years of study at The Hastings Center, we concluded that a greater danger for government manipulation of populations was inherent in low-technology methods—television and preschool education—than from high technological developments—psychosurgery and electrode implantation.[1]

Obviously, there are dangers in any method for modifying human behavior, but the distinction in principle between planting an electrode and implanting an idea is less than one might anticipate. One of the more obvious distinctions is that it is probably easier to withdraw the electrode than the idea. Nonetheless, people are generally more frightened by high technology attacks on their nature than they are by low technology attacks. This disproportionate fear of high technology as a means of influencing human conduct I have labeled the Frankenstein Factor.[2]

Today, we are less concerned about manipulating behavior and more frightened about the prospect of tampering with our genetic structure. Genetic engineering is in the forefront of our anxiety. Genetic engineering also invokes the Frankenstein Factor. We are always unhappy when we are changing our nature. We are concerned that if we fool with Mother Nature we will be made to pay a price. I will tell you in advance that what you are going to hear from me is the good news. I bring glad tidings. I am the optimist in today's crowd. I not only think that we *will* tamper with Mother Nature, I think Mother wants us to.

Spurred by our new potential for altering our species,

questions are being raised about the nature of human nature. Central to this discussion is the concept of human dignity. Human dignity has been under attack from a number of quarters in recent times. We are the despoilers of the environment, the vivisectionists, the polluters, the destroyers. Students are saying, "What a wonderful world this would be if there were only no people in it," and I bristle. The wonderment of the world is a product of human perceptions and human purposes. Dignity has also been under attack in the intellectual world. The groves of academe have been invaded by Tigers and Foxes (and Lorenzes), all devouring our reputation as a species. I propose to tell you how good we really are.

The literature on human dignity is a very small one and surprisingly a relatively modern one. The ancients simply assumed the special position and worth of our species. They had no doubt at all that we were something awesome and wonderful. The ancient Jews, unencumbered by Hellenic concepts of hubris or Christian concepts of humility, saw human beings as supreme among God's creations.

> And God created man in His own image, in the image of God created He them. And God blessed them; and God said unto them: "Be fruitful, and multiply, and replenish the earth, and subdue it; and have dominion over the fish of the sea, and over the fowl of the air, and over every living thing that creepeth upon the earth."[3]

We are not asked to live harmoniously with the nematodes or the viruses of the earth. He commands that we subdue them. In that Biblical passage, we are not asked to be one among creatures, but supreme among creatures, something halfway between animal and God. The special worth of humankind is acknowledged both in our likeness to our Creator, and in His injunctions to us.

The Greeks, despite their concern with hubris, stil acknowledged the special worth of humanity.

The world is full of wonderful things
But none more so than man,
This prodigy who sails before the storm-winds,
Cutting a path across the sea's gray face
Beneath the towering menace of the waves.[4]

For an amazingly long time, the concept of human dignity was founded exclusively on that quotation from Genesis. We were created in God's image and our special worth was vested in that kinship and likeness. This dominated thinking through the Middle Ages.

Macrobius notes that the human race is ennobled by its kinship with the heavenly mind and that of all creatures on earth, only a human shares the mind with the heaven and the stars.

> As far back as written records give us any knowledge, man seems to have considered himself a special kind of being . . . . the biases for medieval views on human dignity were man's creation in the image and likeness of God and his dominion over other creatures.[5]

The early medieval tradition of the image of God was dominated by an Augustinian dualism. There was a clear separation of mind and body. And it was only in the soul that we were in God's image. It was not until the early Renaissance that we first began to develop a concept of the specialness of the species: that something existed in the nature of our species beyond just the affinity of souls that could be interpreted as being in God's image.

For the most part, little movement was made in defining human dignity until modern times, but the Renaissance did produce one important if subtle change. The concept of dignity was gradually extended so that dignity resided not just *in the species*, but in each individual member of the species.

> The problem of human dignity, not in the sense of that of human beings *versus* that of other animals, but in the sense of the

dignity of each human being as a person, has come to the fore with the rise of mercantilism and capitalism and an increasing individual self-awareness. A very clear manifestation of it can be found in the assertion of the protestant reformers that each Christian has to face his God directly and without mediation. The most explicit proclamation of human dignity can perhaps be seen in Kant's second formulation of his "categorical imperative": "Act in such a way that you treat humanity, in your own person as well as in the person of any other, never merely as a means but always also as an end."[6]

The writings of Kant produced the most profound changes in modern philosophy and modern thinking about ourselves. His definition of human worth ultimately became the exclusive definition. Kant defined the special value of our species as residing in our autonomy. The literature of autonomy flourished, and the concept of dignity remained essentially unexamined and neglected.

Dignity was to become an issue in modern writings without ever having been adequately analyzed, and implicitly always carrying its Kantian definition of autonomy. Kant's position was very clear. The dignity, or the worth, of the human being was not based on our special reasoning powers, although he acknowledged that we were quite different from other animals in this way, but was based on our freedom, our autonomy.

Man in the system of nature is a being of slight importance. Although man has, in his reason, something more than they (other animals) and can set his own ends, even this gives him only an *extrinsic* value in terms of his usefulness.

But man regarded as a person—that is, as the subject of morally practical reason—is exalted above any price; for as such he is not to be valued as a mere means to the ends of others or even to his own ends, but as an end in himself. He possesses, in other words, a *dignity* (an absolute inner worth) by which he exacts *respect* for himself from all other rational beings in the world: He can measure himself with every other being of this kind and value himself on a footing of equality with them . . . .Autonomy then is the basis of dignity of human and of every rational nature.[7]

The use of the word dignity and the proliferation of guarantees of dignity in international law and international contracts has become so great in recent years that it demands attention in its own right. I selected one example from the Belmont report—a report of the National Commission on the Protection of Human Subjects. The report starts with an enunciation of basic ethical principles: "Respect for persons incorporates at least two basic ethical convictions: first, that individuals should be treated as autonomous agents, and second, that persons with diminished autonomy are entitled to protection."[8]

The concept of dignity is clearly equated in this report with Kantian autonomy. If we accept the Kantian definition of dignity, human dignity actually starts with the Fall. Adam and Eve did not possess human dignity—grace, perhaps, but not dignity—while living in the Garden of Eden. The dignity of our species starts with the Fall, when they chose freedom and autonomy, with all of its pains, suffering and risk, over security. They chose it and we are stuck with it. We have that risk, but we are elevated because of it. The word sin is never used in relationship to the Fall. The first sin actually mentioned in the Bible is in relationship to Cain's attack on Abel. God created us with the inherent capacity to shape our own existence. That we chose the ambiguities and the dangers of freedom may have distressed Him, but must ultimately have fulfilled His essential design.

Having said all this, what is the problem? We managed to survive with an unexamined concept of dignity conflated and submerged into the principle of human autonomy. The problems started in the late nineteenth century, when psychiatry and psychology began to systematically attack the concept of human autonomy.

Psychiatry and psychology are dominated by two

strong currents. One stemming from Freud, which I will call dynamic psychology, sees behavior as in a dynamic state of tension influenced by unconscious and irrational forces. The other, deriving from Pavlov through Watson to Skinner, is a behavioristic approach. This approach does not acknowledge feelings, emotions or the unconscious, but approaches behavior descriptively. It aspires to being a measurable science.

These two psychologies have little in common except that they intersect at one point. They both agree that behavior is not free. That which is seen as a voluntary choice at any time is actually an inevitable product of inbuilt experiences, prejudices, and determinants.

Freud, in his great discoveries (and I will draw from this psychology, with which I am most familiar), evolves principles which make man less than a rational animal and certainly not a totally free agent. He saw behavior as motivated towards a future goal. He saw it as casually related to the past. He insisted that we did not operate in an actual world, but imposed a world of our own perceptions onto that actual world. Each of us, exposed to similar stimuli, would perceive them differently and therefore respond differently.

The concept of psychic determinism—of a present that is determined by and is a product of the past—this concept that the child is indeed the father of the man wreaked havoc with the law and wreaked havoc with our sense of dignity in ways that were not anticipated by Wordsworth. Under the influence of two giants of modern jurisprudence Harold Lasswell and David Bazelon, the concept of psychic determination was inserted into the law. We got into an absolute mess.

The law demands responsibility and in medicine we do not assign responsibility to an "illness." To a psychiatrist, innocence is an age, and guilt is an emotion. We also operate under the model of the sick role. If you come into my office with a foul abscess on your leg, I won't say, "Get that ugly thing out of my office." I am aware that you are not morally "responsible" for your illness. You are non-culpable. But if you are going to label every antisocial act "sick," every rape, every attack, every murder; if you are going to "explain" each of them with sociological or psychological exculpation, you will eventually destroy the concept of responsibility. That is indeed what is beginning to happen. More important, you are going to fragment the real world, and the common law, which depends on our having a common experience, is going to be distracted by a concept of multiple realities and multiple experiences.

The attack on our concept of ourselves as autonomous agents reached its most frightening and logical conclusion with the publication of B. F. Skinner's *Beyond Freedom and Dignity*. This difficult but important book was widely praised at the time of its publication by many who obviously did not completely comprehend its implications. The use of language in the book is at times confusing. Although "dignity" is used in the title, it is never referred to in the book. Dr. Skinner talks about the "dangerous" concept of human autonomy. I was therefore curious why he used the word "dignity" in his title and how he understood it. I wrote him and asked.

> I used the word "dignity" rather in the French sense as meaning "worth." I was concerned with the credit we give an individual for his achievements and the necessary erosion of that credit as a scientific analysis attributes more and more of a person's achievements to genetic endowment and personal history.
>
> I am very much concerned about the misuse of the term "rights" and I think that "the right to dignity" is particularly objectionable. We defend ourselves against those who would deprive us of the chance to achieve and in doing so, I think, "defend our right to achieve."

> Putting it in a different way, increasing knowledge of the causes of human behavior reduces the role of a supposed free agent to whom we credit behavior. The first reaction is to reject such a science, to preserve freedom and credit for what one has done.[9]

What Dr. Skinner goes on to say in his letter is to confirm the message of his book. He is frightened by our freedom, he is afraid that that freedom is going to destroy us, and he says that it is time to stop acting as free agents. He has the courage of his convictions and I honor him for that. He says if freedom is going to lead us to self-destruction, we do not want that kind of freedom. We had better make sure that we design our descendants in such a way that they are not free to choose evil and destroy us all. Dr. Skinner reassures us, as part of his solution, that we pay a small price. All he asks is that we abandon an illusion; there is no such thing as freedom anyway. We gain security by abandoning a myth.

I had come to know Dr. Skinner and to honor and respect him, and while I maintained my tenuous adherence to the concept of human autonomy, provoked by his arguments I began to question whether there were not other attributes beyond autonomy in which one could vest the special worth of the human being. I think there are dozens. I have selected five unique and inter-related attributes of our species that make us special: Conceptual thought; the capacity for technology; the range of human emotions; a "Lamarckian" genetics; and the last, if not autonomy, is very close to it, and that is freedom from instinctual fixation.

1. Conceptual thought is an exclusively human capacity and is a function of that extraordinary thing called the human brain. I list only four specific manifestations—language, symbolism, anticipation, and imagination. Language is the primary example. A dog can say "whoof, whoof," verbally indicating he wants something, but he

doesn't tell you whether he is in the mood for cheese or biscuits or a walk near the hydrant. In recent times everybody seemed to be training apes to do all sorts of interesting things. The headlines in the science journals and public press informed us that "monkeys can talk." My retort to that was constant. I was really not interested until they said something worth quoting. I was greatly reassured therefore when as distinguished a psychologist as Herbert Terrace, a primary researcher in this area, concluded that primates were not capable of true speech. These days it is, for the most part, the humanist who reduces homo sapiens and the biologist who reveres our species.

Conceptual thought could have arisen only in a multicellular animal, an animal with bilateral symmetry, head and blood system, a vertebrate as against a mollusk or an arthropod, a land vertebrate among vertebrates, a mammal among land vertebrates. Finally it could have arisen only in a mammalian line which was gregarious, which produced one young at birth instead of several, and which has recently become terrestrial after a long period of arboreal life.

There is only one group of animals which fulfills these conditions . . . . Thus, not merely has conceptual thought been evolved only in man, it could not have been evolved except in man.[10]

Language, and what we do with language, is an ennobling aspect of the human being. Cicero based his concept of the uniqueness of man on language.

For there was a time when men wandered at large in the fields like animals and lived on wild fare; they did nothing by the guidance of reason, but relied chiefly on physical strength; although their ignorance and error, blind and unreasoning passion satisfied itself by misuse of bodily strength, which is a very dangerous service.

At this juncture a man, a great and wise one I am sure, became aware of the power latent in man . . . . He introduced them to every useful and honourable occupation, though they cried out against it at first because of its severity, and then when through reason and eloquence they had listened with greater attention, he transformed them from wild savages into a kind and gentle folk.[11]

The next aspect of conceptual thought needs very little elaboration. The use of symbolism allows us to do all sorts of extended things. The world of mathematics rests on our capacity to abstract and symbolize. Poetry and the arts are testaments to its power.

Anticipation utilizes intellect, memory, and abstraction (symbol formation). It represents a great advance in the struggle for survival. The simplest creatures, the amoebas, have no sophisticated mechanism of survival. They ingest certain things and if nonnutrient eject them. As you move up the biological ladder you find animals that have a tropism towards nourishing things and away from painful things. You can see in this the design of God if you wish, or you can simply see Darwinian selection. It is possible that a mutant occurred at one time that loved poisons and fire, and hated veggies and the other things on which one survives. Such a mutant would have lasted barely one generation.

The next advance in survival mechanisms was the development of the primary emotions—the stress emotions of fear and rage—which we share with many lower animals. If you depend on pain alone, you know that you are in danger only when the alligator presses his teeth into your flesh. At that point it may be too late to appreciate that you are in a threatening situation. If you have an appropriate sense of fear then you can anticipate the pain at the mere sight of the predator, and run like hell. In order to anticipate the pain you have to have something called distance receptors—touch alone will not do it. You have to have a sense of smell, hearing, or vision, which allows for the simple anticipation that causes the fear and rage responses that mobilize for fight or flight. With distance receptors and primary emotions you do not have to wait for the alligator's bite. You can see him in the distance, be frightened, and run. If, beyond these, you have imagination and anticipation, you do not even have to wait to see the alligator. You can know that you should not go walking in the swamps at night.

Anticipation can also cause problems. While one can anticipate that which is going to come, one can also anticipate that which is never going to come, like the cloned Hitler soldiers that have been introduced into discussions at this conference. The price we pay for freedom and anticipation is the possibility of creating terrors for our own pain or amusement.

Imagination is a specifically human quality that allows us to experience, as Dr. Lebacqz was suggesting, a feeling that transcends rationality; a learning that comes from love and relationships. It is the basis for all art and culture. There is a quotation from Ovid as he deals with the Prometheus legend in his *Metamorphosis* that sets the imagination at the center of human uniqueness. The Prometheus legend is of interest in defining our species. There are two independent legends. The sin of Prometheus according to one tradition was that he created the human species. His sin in the other legend was that he gave fire, which is usually seen as a symbol of technology, to our species. The answer may be that technology shapes and defines our humanhood to such a great degree that the technological animal becomes the human animal—two different versions of the same concept. Ovid, in describing our creation, says Prometheus, the son of Hepetheus, mixed matter with fresh running water and molded us into the form of the all-controlling God: "And, though all other animals are prone, and fix their gaze upon the earth, he gave to man an uplifted face and bade him stand erect and turn his eyes to heaven."[12]

This poetic expression of Ovid's describes the aspiring nature of our species.

2. The capacity for technology. Dr. Thomas, in response to what he felt was an attack on science by Dr. Lebacqz, said that it was not science that was the culprit, but technology. So, in defending science, he derogated technology. He further said that real facts and real truth and real knowledge never harmed anybody. As a psychoanalyst I have almost the opposite belief. It is the real knowledge—that you are about to die, that you are an inadequate person in relationship to those around you, that your wife was entitled to divorce you because you are a mean, nasty person—it is this real knowledge that is often the most painful and destructive. But, then, I am generally not much concerned with real knowledge. The ultimate value of real knowledge is, for the most part, in its (often unanticipated) applications. The theoretical is the plaything of the creators until applicability lends it nobility through service to humankind. Science is the servant of technology, not the other way around. Technology and application lend glory to science. Current attacks on immoral technologies, such as the current tizzy about recombinant DNA, ignore the fact that technology (with the possible exceptions of instruments of war and torture) is morally neutral. Technologies can be used immorally, but then so can love and food and charity.

3. The third special quality of the human species is the range of our emotions.[13] While lower animals share with us rage and fear, they share none of the ennobling emotions. My wife is convinced that when our dog does some outrageous canine feat he is terribly guilty and he should not be punished. That dog is never guilty. What she is seeing is fear, albeit a "guilty" fear. He is terrified that he is going to be caught. Guilt is an entirely different emotion. If you are going sixty-five miles an hour in a forty-mile-an-hour speed zone and you hear the policeman's si-

ren, the emotion you are feeling is guilty fear. The test of this is the following: If that patrolman passes you by and arrests the speedster in the Porsche in front of you and you feel relief, that is guilty fear. If you feel disappointment, that is true guilt. Guilt is a uniquely human emotion that demands expiation for wrongdoing. It implies that we have internalized a set of noble standards by which we judge ourselves and that if we have done wrong we punish ourselves. In one outrageous "self-help" book, there is a chapter called "Guilt and Anxiety: The Useless Emotions." Since anxiety is the principal emotion serving individual survival, by eliminating these the author has eliminated two of the foundation struts on which our survival is built.

I know inappropriate and excessive guilt causes much pain. However, I am equally concerned with young people who have managed to exorcise all guilt and demonstrate it by knocking little old ladies over the head with lead pipes to collect the four dollars and fifty-seven cents that is in their purses. Those are the guiltless individuals. I do not think that society is better off for having them with us.

4. That we are "Lamarckian" animals cuts directly to our concerns about genetic engineering. There was a great debate some seventy years ago about how the genetic mechanisms of Darwinian evolution worked. There were two theories. One was Mendel's theory, which happened to be brilliant, and happened to be true. The opposing Lamarckian theory was environmentally linked and insisted that acquired characteristics were transmittable. The giraffe, by stretching its neck for leaves, acquires a longer neck and transmits that characteristic to its offspring.

Mendelian genetics have been universally accepted, except for the obtuse period when Lysenko dominated Soviet genetics. Russian agriculture has barely yet recovered from this insistence that science conform to political bias. In the human species alone acquired characteristics are transmitted, not by protoplasmic design, but by the power of a transmitted culture. The power of being able to transmit knowledge to avoid rediscovery is of enormous adaptive value. All of the things that an earthworm or a bird must know to survive are genetically fixed. Since they do not have to learn anything to survive, they learn very little. We are not genetically fixed. Since we do not have to rediscover the wheel anew in each generation, we are free to build on that knowledge to develop the cart and then the car and then the spaceship—and then? The power of culture as an adaptive mechanism was expressed most eloquently by Theodosius Dobzhansky, one of the seminal thinkers in modern biology:

> Our genes determine our ability to learn a language or languages but they do not determine just what is said. The structure of neither the vocal cords nor the brain cells could explain the difference between the speeches of Billy Graham and Julian Huxley.
>
> Culture is not just another mechanism of adaptation; it is vastly superior to the biological mechanisms which spawned it. It is more rapid and efficient. When genes are changed through mutation, the change is transmitted solely to the specific offspring—and only with generations of time enters into the species at large. Changed culture, on the other hand, may be transmitted to anybody regardless of biological parentage, or borrowed ready-made from other people. In producing the genetic basis of culture, biological evolution has transcended itself—it has produced the superorganic. In other words, the kind of brain capable of conceptual reasoning is not only the product of a certain development but is capable of dictating a future development.[14]

5. The final point of special human distinction that I will discuss is our freedom from instinctual fixation. I

would make the point that if we are not truly autonomous agents, freedom from instinctual fixation is as close to autonomy as is necessary to insure our dignity. In two quotations of which I am particularly fond you see how early scholars appreciated the power, the privilege, and the *rightness* of our changing our nature. It is not tampering, but *obligation* to modify ourselves. The Talmud raises the question: "If God had intended man to be circumcised, why wouldn't he have created him that way in the first place?" And these medieval Jewish scholars answered: "God created man alone among creatures incomplete, with the capacity and the privilege of sharing with his Creator in his own design."

This principle was beautifully elaborated by that great Renaissance thinker, Pico della Mirandola.

> Neither a fixed abode nor a form that is thine alone nor any function particular to thyself have I given thee, Adam, to the end that according to thy longing and according to thy judgment thou mayest have and possess what abode, what form, and what functions thou thyself shalt desire. The nature of all other things is limited and constrained within the bounds of law prescribed by Us. Thou, constrained by no limits, in accordance with thine own free will, in whose hands We have placed thee, shalt ordain for thyself the limits of thy nature.[15]

Now, having said all of this, isn't it a bit anthropocentric? Isn't this argument about the superiority of our species biased by the fact that it was written by a human being instead of a mole? Couldn't Ovid be turned upside down? Could not the mole say, "Look, human beings have their heads up there in the clouds and airy spaces. We are at the fundaments of truth, at the roots of the matter, down here under the earth." That is obviously true, and a mole could argue thus. But a mole does not argue, and that is not beside the point. When an argumentative mole appears I will be happy to get involved in the first debate. But so far he has not and I am safe. The reason I will be safe in the foreseeable future is expressed by another great biologist, George Gaylord Simpson.

> Even when viewed within the framework of the animal kingdom and judged by criteria of progress applicable to that kingdom as a whole and not peculiar to man, man is thus the highest animal. It has often been remarked. . . . that, if, say, a fish were a student of evolution he would laugh at such pretensions on the part of an animal that is so clumsy in the water and that lacks such features of perfection as gills or a homocercal caudal fin. I suspect that the fish's reaction would be, instead, to marvel that there are men who question the fact that man is the highest animal. It is not beside the point to add that the "fish" that made such judgments would have to be a man.
>
> Is it necessary to insist further on the validity of the anthropocentric point of view, which many scientists and philosophers affect to despise? Man *is* the highest animal. The fact that he alone is capable of making such a judgment is in itself part of the evidence that this decision is correct.[16]

We are considering one of the most profound and radical discoveries of modern biology. Recombinant DNA technology opens awesome capacities for the future. Of course this technology is subject to abuse, like almost all others. All powers have the potential for corruption. We have in the past used some of our powers unwisely and unwell. There is I believe less to fear here than the prophets of doom would have you believe.

The capacity to do evil, while a risk of freedom, is a component in defining the good. I center my hopes on that freedom. While I am not so foolish as to discount the fears of some of the other panelists, I recognize that having these is the price we have always paid for freedom. But I also know that freedom is intrinsic to what makes our species worthwhile. If we sacrifice that freedom, if we sacrifice that special inquiry, that capacity to look to the stars, to reach out for something new, we cease to be a species particularly worth saving. The human species is the glory of creation. I have very little sympathy for people who talk of the "rights" of animals. While I feel that human dignity demands that we treat animals with a certain amount of worth, only human beings have rights. While I may devote energies to the preservation of the Sequoias, I do so with no misguided claim that trees have standing. I have pointed out to students who affect a contempt for people, while esteeming nature, that the sun does not even "set" except in the perception of our minds. I am in good company here. Whitehead has said:

> Thus nature gets credit which should in truth be reserved for ourselves: the rose for its scent; the nightingale for his song; and the sun for his radiance. The poets are entirely mistaken. They should address their lyrics to themselves, and should turn them into odes of self-congratulations on the excellency of the human mind.[17]

"Respect" for lower creatures is often a veiled attack on homo sapiens, an attempt to compromise an essential dignity that has been recognized since the beginnings of culture.

I will close with a quotation from William James that sums up both my anthropocentrism and my faith in the moral nature of human relationships:

> We have learned what the words "good," "bad," and "obligation" severally mean. They mean no absolute natures, independent of personal support. They are objects of feeling and desire, which have no foothold or anchors in Being, apart from the existence of actually living minds.
>
> Wherever such minds exist, with judgments of good and ill, and demands upon one another, there is an ethical world in its essential features. Were all other things, gods and men and starry heavens, blotted out from the universe, and were there left but one rock with two loving souls upon it, that rock would have as thoroughly moral a constitution as any possible world which the eternities and immensities could harbor. It would

be a tragic constitution, because the rock's inhabitants would die. But while they lived, there would be real good things and real bad things in the universe, there would be obligations, claims, and expectations; obediences, refusals, and disappointments; compunctions and longings for harmony to come again, and inward peace of conscience when it was restored; there would, in short, be a moral life, whose active energy would have no limit but the intensity of interest in each other with which the hero and the heroine might be endowed.[18]

NOTES

1. Willard Gaylin, Joel S. Meister, and Robert C. Neville, eds. *Operating on the Mind: The Psychosurgery Conflict* (New York: Basic Books, 1975).
2. Willard Gaylin, "The Frankenstein Factor," *New England Journal of Medicine*, 22 September, 1977.
3. Gen. 1:27–28.
4. Sophocles, "Antigone" in *Oedipus the King*, ed. and trans. Peter D. Arnott. Crofts Classics Series, lines 327–331.
5. Richard C. Dales, "A Medieval View of Human Dignity." *Journal of the History of Ideas* 38, no. 4 (October–December 1979): 557–59.
6. Axel Stern, "On Value and Human Dignity," *Listening*, Spring 1975, p. 78.
7. Immanuel Kant, *The Doctrine of Virtue*, trans. Mary J. Gregor (New York: Harper Torchbooks), p. 99.
8. *The Belmont Report*, DHEW Publication No. (OS) 78–0012 (September 1978), pp. 4–5.
9. B. F. Skinner, in a personal letter to Willard Gaylin, 18 September, 1978.
10. Julian Huxley, *Man in the Modern World* (New York: Mentor Books [NAL], 1944), pp. 16–17.
11. Cicero *De Inventione*, trans. H. M. Hubbell, The Loeb Classic Library 1.1; 2.2, p. 5.
12. Ovid *Metamorphoses*, trans. Frank J. Miller, The Loeb Classic Library 1.7.
13. For a full discussion of the range of human emotions, see Willard Gaylin, *Feelings: Our Vital Signs* (New York: Harper & Row, 1979).
14. Theodosius Dobzhansky, *Mankind Evolving* (New Haven: Yale University Press, 1962), pp. 346–47.
15. Pico della Mirandola, "Oration on the Dignity of Man" in Ernst Cassirer et al., *The Renaissance Philosophy of Man* (Chicago: University of Chicago Press, 1956), p. 224.

16. George Gaylord Simpson, *The Meaning of Evolution*, The New American Library (New York: Mentor Books, 1956), p. 139.
17. Alfred North Whitehead, *Science and the Modern World* (New York: Free Press, 1967).
18. William James, *The Writings of William James* ed. J. McDermott (New York, Random House, 1967), pp. 618–19.

Manipulating Life: The God-Satan Ratio

CLIFFORD GROBSTEIN

This conference is considering implications of manipulating life, particularly human life: manipulation that may arise out of advances of biomedical knowledge. The conference title calls special attention to our "human responsibilities" in connection with such manipulation or intervention. I shall focus on new and prospective reproductive interventions that have been the subject of recent agitated debate. The range of prospective reproductive interventions is far wider than can be covered in this presentation. I have decided, therefore, to concentrate on external human fertilization, gene transfer, and a possible intersection between the two that has generated special concern. These matters have come up in earlier presentations at this conference. Let me briefly recapitulate what each refers to.

External human fertilization is often referred to as *in vitro* fertilization (IVF), meaning that the external process is carried out in laboratory dishes (sometimes but not necessarily glass). Egg and sperm, obtained from a couple unable to conceive naturally, are brought together under carefully controlled laboratory conditions. Fertilization and subsequent development of the egg are confirmed microscopically and, in the standard medical application, the early embryo is transferred back to the uterus of the egg donor. In more than one in ten clinically reported attempts, pregnancy and normal birth ensue. It is reasonably estimated that well over one hundred and fifty babies now have been born around the world as a result of the procedure.

Gene transfer, on the other hand, is not known to have been successfully applied medically but its clinical potential has been widely discussed. The potential is the outcome of the past decade of progress in molecular genetics. DNA segments containing a single gene, plus its control elements, can be isolated and inserted into laboratory-cultured cells. Under suitable circumstances, still subject to further clarification, the inserted genetic material is combined with that of the recipient cell and passed on to daughter cells. The incorporated gene can be expressed in the formation of its normal products—it introduces new hereditary properties into the recipient cell. In a few instances, such genetically altered cells have been returned to whole animals of the species of origin—thus conferring on these animals the altered cellular traits.

The clinical applications most widely seen for such gene transfer of hereditary defects is to sickle-cell anemia and other diseases caused by abnormal hemoglobin. Afflicted individuals have two defective hemoglobin genes in their blood-forming cells. Since normal human hemoglobin genes can be isolated and transferred, the idea would be to introduce them into the blood-forming cells of a patient's bone marrow. If successful, the modified bone marrow cells would give rise to normal blood cells rather than defective ones, hopefully curing the disease through "gene therapy."

Were this to work, it would cure the disease in the afflicted individual but would not change the chances of passing the disease on to offspring. This is because the cells that normally give rise to eggs and sperm become a segregated cellular lineage very early in development. Changing the heredity of bone marrow cells thus does not itself alter heredity in the cells that give rise to a new generation. If the objective is fully to eliminate the disease not only in an afflicted individual but in descendants, the change must be made in germ-line cells that give rise to gametes.

This is where the paths of IVF and gene transfer may intersect. In the IVF process early human embryos are, for the first time, accessible to observation and intervention. In fact, mouse embryos of a similar early stage have been subjected to gene transfer and, in very early and still limited results, the transferred gene has been shown to be present and working not only in the adult mice derived from the treated embryos but in a small percentage of the generation they give rise to. The objective of this line of research in mice is not clinical application but the production of mouse strains in which gene operation and control can be more effectively studied. Nonetheless, if fully validated and raised to the levels of effectiveness and safety demanded for human application, these efforts theoretically could lead to a first direct human hereditary intervention in the inter-generational sense.

Commentators on these developments and their implications have frequently raised the question whether the physicians and scientists who are involved are "playing God." This is not only because of possible conflict with particular religious teachings but because of the unprecedented aspects of the interventions in previously inaccessible and little understood processes. I share with these commentators their sense of the profundity of the issues. However, I believe that the image conveyed by invoking the concept of playing God is not helpful as a guide to the solution of the serious policy questions raised. It exacerbates differences among us and makes more difficult the pluralistic accommodation that is essential in a society that respects and protects religious and other individual freedoms but seeks consensus in its public policy.

Before saying more about the topic itself I should like to recount some personal history that bears on my own perspective with respect to it. I have been involved in three of the major scientific, medical, and policy currents of this century. The first is the explosive development of the biomedical sciences, of which the new potential for human reproductive interventions is part. My activities as a "bench scientist" were dominated by this biomedical explosion from the time, shortly after the Second World War, when I joined the staff of the National Cancer Institute as one of the very early NIH postdoctoral research fellows.

The second major current is the development of nuclear weaponry based upon advances in atomic physics. Nuclear weaponry was the subject of bitter controversy in Washington circles when I arrived in suburban Bethesda in the late forties. I was fresh from helping train combat crews for B–29 bombers, the delivery vehicles for the two primitive fission bombs that devastated Hiroshima and Nagasaki. My military station for much of the preceding two years was within easy drive of Los Alamos and Alamogordo—of whose operations I knew little at the time. However, in postwar Washington I found myself on the fringes of the climactic policy debate over whether to develop the "Super," the hydrogen bomb. Like many scientists of my generation, whether or not participants in actual weapons development, I regard the Bomb as a constant and overwhelming moral and political challenge to scientists and nonscientists alike. Nothing else rivals in importance this challenge to national and international security.

The third current is the tightening circle of interaction among science, technology, and public policy, an interest of mine that antedated the Second World War and has become my focal academic activity in the past decade. The sharply enhanced capability for purposeful manipulation of human reproduction is but the latest of the dilemmas presented to public policy by advancing scientific knowledge and derivative technology. Only through the understanding of the close linkages among science, technology, and policy can measures be found to meet our human responsibilities in manipulating life. I want to examine the concept of "playing God" from this policy perspective, beginning with the meaning it appears to have for those who use it as an expression of concern.

What is referred to, of course, is the scope of the human role, relative to the role of God, in governing the human experience. Conceptions of the nature of God's role vary widely, both over cultural history, among existing religions, and in secular society. The common religious assumption is that the human role is far more limited than that of God, with a distinct delimitation between the two. Human activities that penetrate the boundary are not only sacrilegious in themselves but pose a risk of bringing down unanticipated and unwanted consequences. Thus, the general secretaries of the National Council of Churches, the Synagogue Council of America and the United States Catholic Council, in a joint letter to the President on June 20, 1980, said, "We are rapidly moving into a new era of fundamental danger triggered by the rapid growth of genetic engineering. . . . New life forms may have dramatic potential for improving human life, whether by curing diseases, correcting genetic deficiencies, or swallowing oil slicks. They may also have unforeseen ramifications, and at times the cure may be worse than the original problem—history has shown us that there will always be those who believe it appropriate to 'correct' our mental and social structures by genetic means, so as to fit their vision of humanity. This becomes more dangerous when the basic tools to do so are finally at hand. Those who would play God will be tempted as never before."

The intent of the general secretaries was to have their concern explored within the federal government. Two years later the President's Commission for the Study of Ethical Problems responded in its report entitled "Splicing Life." The Commission noted that "while religious leaders present theological bases for their concerns, essentially the same concerns have been raised . . . by many thoughtful secular observers. . . . The examination of the various specific concerns need not be limited, therefore, to the religious format in which some of the issues have been raised."

Having examined the several meanings that might be attached to the notion of playing God, the Commission concluded that none justified prohibition of continued research in the area but that the issues are consequential enough to warrant a mechanism for continuing oversight. Several months later, however, a broad group of prominent religious leaders specifically called for prohibition of genetic transfer into human germ-line cells, transfers that might have consequences in future generations.

What clearly continues to haunt us is the issue of Genesis or Creation: Bible-based religious groups generally view God as the Creator of the Natural Order. Humans are part of the Creation but are transgressing when they seek to intervene in the inner workings of the Natural Order. In particular they are transgressing when they intervene in human beginnings, nature, or destiny because these are matters of God's special concern.

In fact, of course, there has been steady nibbling at the edges of this religiously defined sacred province. People have been meddling in natural processes since they became people, meaning that the human sphere for intervention in the natural order has been enlarging relative to the divine one. Along the way, there have been recurring anxiety and controversy, with repeated admonitions and dire prophecies from the protectors of the divine role that each new transgression goes too far. Nor have the warnings been without support in experience. Many human interventions in the natural order—undertaken with the most beneficent intentions—have had unanticipated unfortunate consequences. There has been much dispute in recent years over the net gain of our environmental, health, and even economic interventions. Many voices call for a respite, for time to draw breath, to evaluate, to adjust to what already has been wrought, even permanently to let well enough alone.

It is in this context that the charge is heard that scientists are now finding new ways to "play God" and, at this very time, in some of the most deeply sacred and private processes of human existence. The charge was heard in the middle seventies when expanding knowledge of the nature and hereditary role of molecular DNA gave clear indication that a potential was being created to alter the heredity of human cells. It was repeated at the turn of the decade when the clinical success of external human fertilization, itself an unprecedented intervention in a previously sacrosanct area, opened a window for still further interventions in human development. Between the two a model was presented in laboratory animals for limited alteration of succeeding generations. The advent of this potential new age of human intervention seemed close to the ultimate in arrogant encroachment on divine prerogative. It was this that led the general secretaries to call upon the president of the United States to provide careful assessment of the morality of new biomedical technologies that might be on the horizon.

Clearly, a charge of playing God is less heinous than a charge of playing Satan. To play God is, by intention at least, to seek benefit. It would not be God-like to be malevolent. Foolish arrogance, not malevolence, is what biomedical scientists are charged with displaying. But what of their colleagues, the physicists, who produced the atomic bomb in the forties with full understanding that the purpose of the effort was military? Were they not playing Satan? J. R. Oppenheimer himself acknowledged that the physicists had come to know sin. Was the whole episode not typically satanic, with its early rationale of benefiting mankind by beating the evil Hitler to the punch—followed, after the collapse of Hitler without resort to atomic bombs, by the unleashing of the holocaust upon the Japanese? Is the scenario not now obviously satanic when possessors of nuclear weapons, weapons grown mightier and more numerous, hold the entire world hostage on a knife edge of mutual deterrence? Were the physicists of mid-century not playing Satan?

At some point, however, one must ask whether such dramatic but simplistic portrayal of complex social decision-making really is useful. Albert Einstein and Leo Szilard played key roles in getting the attention of President Franklin D. Roosevelt, an essential step to the launching of the Manhattan project. Both were peace-loving men who labored mightily after Hiroshima and Nagasaki to constrain further development and use of nuclear weapons. What do we gain by regarding such men first as agents of Satan and then of God? Is it not more to the point to say that they were human beings seeing the matter first in one context and then in another as circumstances changed? Did the travail of J. R. Oppenheimer stem from his being a double agent of God and Satan or from the conflicted structure of his own ambiguous, self-chosen role? What understanding of any of these people or the great issues they faced is provided by portraying them in the rhetoric of a medieval morality play? Do we not strike closer to human responsibilities by facing as our own the choices between moral alternatives, casting aside masks culturally created to displace responsibility to some benign or malign higher authority and power? Is it not time, past time, to accept that it is we who should improve the God/Satan ratio of our own behavior?

The central fact is that our growing options and powers have increased to levels that force attention to human responsibilities once assigned to the gods. *We* as human beings have altered the face of our planet, *we* unleashed and in some measure now control the primordial power of stars, *we* have dispatched vehicles and messages beyond the solar system, and we may now be approaching capability to manipulate the substrate of our own genesis. What hangs over us as a new millennium nears is not so much sacrilege as anxiety, the dread that we (more especially our fellows) may not be able to match knowledge with wisdom. It is, indeed, an epochal challenge.

I submit that a calm, retrospective look at the events of the past decade provides some evidence that we are beginning to beat out a path to wisdom. For example, we have moved through the acute anxiety attack that followed the advent of DNA recombination in the early seventies. That anxiety fixed on biohazard, the possibility of generating new plagues and pestilences through inadvertent release of altered organisms from the laboratory. Thanks to the principled approach of the scientists most directly involved, risk-mitigating measures were adopted that allowed for political and social adjustment to the new realities. The process is still going on, with broader and broader issues that earlier were avoided now being seriously addressed. There is no better example of this broadening than the matter of possible deliberate clinical application of gene transfer.

How difficult is this issue in terms of appropriate policy? It falls into two parts with a transitional or gray zone between. The first part involves gene transfer to offset genetic defect in somatic cells, the scenario earlier outlined to deal with sickle-cell anemia and other genetic defects affecting hemoglobin. An abortive clinical effort in this direction actually was made in 1980, unsuccessfully, but demonstrating that policy currently exists to monitor and, in some degree, to regulate advance along these lines. Two separately developed policies were invoked, the first governing experimentation on human subjects, the second governing experiments involving recombinant DNA. At the University of California at Los Angeles a faculty member sought approval from the institutional committees that must review proposals in these areas. The proposal was to remove bone marrow cells from patients severely ill due to defect of their hemoglobin gene, to treat the bone marrow cells externally with normal hemoglobin genes that might be produced by recombinant techniques, and to return the hopefully normal bone marrow cells to the patient to produce normal red blood cells.

When there was delay in gaining approval from UCLA committees the investigator elected to attempt the procedure abroad where approval had been given. When his efforts involving two patients in Israel and Italy became known, both UCLA and the NIH, following investigation, reprimanded and punished the researcher for violation of their required procedures. At issue, in the background, was the question whether therapeutic modification of human bone marrow cells was any different in policy terms if carried out genetically rather than by chemotherapy or hormone replacement.

The informal consensus of commentators on the UCLA case is that gene transfer is not fundamentally different from other forms of therapeutic manipulation, assuming that the change benefits the individual patient and the altered cells terminate with the life of the treated individual. If this consensus holds, gene therapy is likely to be sanctioned at the appropriate time on a case-by-case basis under existing policy if rules for clinical trials are followed, including demonstration of probable safety and efficacy in animal tests. Although no further attempts at clinical application have yet been reported, several lines of laboratory investigation point toward achievement of necessary preconditions for clinical trials, very likely by the end of this decade.

Evaluation and choice are distinctly more complicated for modification of human germ-line cells. Here we do face difficult questions and need to proceed with caution while we deliberate. Is there any convincing rationale for interventions in human beings if effects will persist into future generations? So far as consensus has yet been achieved the following rationale might be acceptable. Patients who are afflicted with a heritable hemoglobin defect are not only sick in their blood-forming cells but also in their gamete-producing cells. Their blood cell defect threatens the duration and quality of their own life, but their germ-line defect reduces their reproductive health by increasing the risk of defective offspring. If their somatic defect were corrected the state of their germ cells would no longer be academic if they could produce children. Seen from the point of view of the individual patient, therefore, there appears to be as good rationale for treating the germ-line cells as the blood-forming cells.

It can be argued against this patient-oriented view that subsequent generations are being simultaneously altered without consent. But if the change is regarded as beneficial to this generation, assuring them effective rather than defective hemoglobin, why should the next generation not be equally accepting? Under our existing medical model it can be argued persuasively that genetic modification of germ-line cells should be sanctioned if it is oriented toward patient benefit and poses no serious harm to offspring.

Given successful application along these lines, however, would there be pressure to expand the sanctioned purposes? Might it be suggested, for example, that not only individuals expressing the hemoglobin defect should be treated but recessive carriers as well? Imagine a case of a couple who already has had a sickle-cell or thalassemic child. Both carry the same recessive defect and they therefore are at 25 percent risk of producing an offspring with the actual disease. If either were successfully treated the couple's risk of having an actually defective offspring would drop to zero, although half of their children might be carriers.

Assuming that the treatment is fully effective and at zero risk (a very speculative assumption in terms of current knowledge), treatment of one member of the couple fits the model of preventive medicine—intervention to reduce disease incidence, as with vaccination, but in a succeeding generation. Treatment of both members of the couple would further reduce actual disease incidence, but not until the second generation—the couple's grandchildren. The additive effect of treating both members of the couple, in terms of preventive medicine, is quantitatively small and two generations removed. However, in terms of genetic medicine, including what is called negative eugenics, the cumulative effect of treating all recessive carriers over time could be elimination of the recessive defect from the population. This, however, would require a deliberate and sustained effort based on a formulated purpose that goes beyond the conventional medical model. It no longer is confined to benefit for an individual patient or couple and immediately succeeding offspring. It is motivated by concern about the human gene pool.

Thus, in considering treatment of germ-line cells, there is a gray zone containing a possible divide for gene-transfer policy. On one side of the divide there is medical motivation to improve the gene pool. Between is a zone of overlapping motivation in which it is difficult to draw a sharp line without reference to intent or purpose. A comparable gray zone exists for external human fertilization. As a means of overcoming sterility in married couples IVF now appears to have been accepted under the umbrella of previously developed biomedical policy—to provide benefit to afflicted individuals or couples who cannot naturally produce a child. But the issue of freezing early IVF embryos has revealed another possible divide. Embryos have been frozen at some treatment centers when more embryos are obtained from a hormonally treated patient than are usefully transferred to the uterus of that patient in a single attempt to achieve pregnancy. In several instances such frozen embryos have been thawed and observed to be capable of continued development *in vitro*. In one instance a thawed embryo continued development for several months on transfer to the uterus of the donor, although the pregnancy terminated prematurely. The primary medical rationale for freezing early embryos thus is to avoid their destruction and to give time for transfer back to a uterus for continued development.

To transfer frozen-thawed embryos to the uterus of the egg donor in subsequent cycles is a technical extension of IVF that provides benefit to the donor-patient. Embryos derived from a single hormonally stimulated cycle can be used for successive transfer attempts without additional laparoscopy, the most uncomfortable step for the patient. This use of freezing falls therefore within the medical model for improved patient care. However, two possible variants in the use of frozen embryos illustrate a gray zone that strains the medical model and even goes beyond it altogether.

The first variant is to use surplus embryos, after meeting the needs of the donor couple, for transfer to another sterile couple. Such "embryo adoption" by the second couple is technically possible, and pregnancy following transfer of an embryo from one woman to another has been recorded. The procedure is routine in cattle. It remains within the medical model in that it is an effort to overcome sterility, but raises unusual social and legal issues about family relationships of the offspring that have brought expressions of concern from a number of commentators.

A second imaginable variant that is not known to have been attempted but seems technically feasible involves freezing of IVF embryos early in a marriage with delay of transfer to the donor uterus until some years later. It has been noted that many women are postponing their pregnancy in their twenties for career purposes but seeking it in the middle to late thirties or even in the early forties. By the natural process such late pregnancies are more difficult to achieve than early ones and have a higher incidence of genetic abnormality—presumably due to aging of ova in the ovary. The imagined variant of banking ova or embryos early in life might not only circumvent the aging process but also provide full control of family planning. Such options lie at the edge or beyond the boundary of the conventional medical model. They also clearly involve power in human decision-making that exceeds traditional religious conceptions. What is our human responsibility in confronting such consequential technological options?

Human beings have long been meddlers in the natural order but they have also been contemplators and moralizers. Clearly the new reproductive options call for deliberative contemplation and possibly for new moral principles. The experimental excesses of German Nazism brought the Nuremberg code emphasizing individual human rights that has ever since strongly influenced clinical trials around the world. It is our human responsibility to formulate new principles, if possible, *before* rather than *after* excesses occur.

The process has, in fact, already begun: witness the voluntary scientific moratorium on certain kinds of experiments established by molecular geneticists in the mid-seventies, the study and report on external human fertilization of the Ethics Advisory Board of the Department of Health, Education and Welfare, and the recent recommendations of the President's Commission on Ethical Problems on "Splicing Life." One can discern in these and related activities both here and abroad an innovative mode of public deliberation to meet the new policy—the challenges we face. We are in a period of improvisation and invention in mechanisms of policy-formation to match the thrust of our scientific and technological advances.

If the process of deliberation has begun, what further steps are needed? What has characterized the new approach so far is, first, an attempt to ensure full involvement of all relevant interests and, second, emphasis on ethical considerations. Wide involvement of interests is important because the issues raised may bear on the nature and future of humanity, obviously of concern to all human beings. In turn, this broad context focuses attention on values and on concepts of human nature and purpose—the crux of ethical discourse.

If these considerations are to be incorporated in a policy-forming process for IVF and gene transfer, there must be the widest possible awareness and discussion of the new anticipated options and their implications. This conference is a step in the right direction but it needs to be magnfied perhaps a thousandfold to accomplish the purpose. Particularly important to the consciousness-raising effort is participation by young people who will make the individual and collective decisions and raise the resulting children. College and university faculties accordingly must recognize a new agenda, not only for teaching but for scholarly effort to clarify the knotty ethical and social dilemmas that are posed.

Beyond this there is a need for continuing interactive public forums, hopefully of more than one kind. Most often suggested is a national commission along the lines of the Ethics Advisory Board (EAB), the National Commission on Research on Human Subjects, and the President's Commission for the Study of Ethical Problems. A new commission to exercise oversight on genetic intervention has been proposed in legislation introduced by Congressman Albert Gore and passed by the House. The proposal does not include developmental interventions such as IVF; it would be desirable for it to do so. But more important than questions of scope is the basic role of such a commission or of other forums that may be established to deal with the subjects under discussion.

I suggest, as have many others who have addressed the matter, that what currently is needed is a deliberative forum and not a regulatory body. Definitive decisions and their implementation are not the prime objectives. Clearer perspective, greater understanding, and wider public awareness are first needed. The commission should generate national discussion, with broad participation, through public hearings, commissioned studies, and widely distributed progress reports. Its short-term objective should be to scout the territory, to develop priorities for deliberation, and to identify and disseminate seminal thinking. This was also the role of the EAB and the Presidential Commission. The effort needs to be continued until wider consensus on approaches is achieved.

What conclusions may eventually come from such a commission and from other deliberative forums? Two general approaches have been discussed for decision-making in areas as complex as those we are discussing: (1) consideration case by case and (2) decision by principle. In most extreme form case-by-case conclusion implies that each case is unique, no precedent or generalization exists before or is established as each case is decided. Decision on principle, however, implies that generalizations—developed either *a priori* or out of tradition or synthesis of what is common to previous cases—are guides or even determinants of decisions about each new case. In many practical situations both models actually are used. A combination seems desirable in the present circumstances, since many novel cases are likely to be presented that may not have obvious precedents or other applicable principles but, on the other hand, rejection of all principles may increase anxiety among those who approach these matters entirely as a matter of traditional morality.

What sort of principles might be considered for early adoption—possibly to be tested and amplified as specific cases are considered and decisions reached? Principles might be adopted early and provisionally because they are believed to command general support and because the clarification achieved by their adoption may reduce anxiety and controversy. Several examples may be suggested that fit this category. The first is an extension of existing principles of human rights that include assurance that individuals will not be subjected to experimentation without informed consent. Such a principle might provide that "no genetic or developmental intervention will be practiced on any human being that is intended, or can reasonably be expected, to restrict or reduce that individual's biological capability, either within the individual's generation or in the production of offspring." The objective of the principle is to preclude such uses of genetic transfer as might create subhumans or castes as a basis for a hierarchical authoritarian society. Although this is technically an unlikely scenario, according to current knowledge, it has been widely discussed and generates much apprehension. To set it formally outside the bounds of acceptability would reduce the dread that for many people colors the entire subject.

The second example also addresses an area of public concern but is more difficult to deal with categorically. It relates to human germ-cell modification, which we already have seen involves a gray zone between it and somatic cell modification and raises the complex issue of eugenics. The following principle provides for special oversight in this area. Its exact language is likely to be the subject of much debate as it is addressed in detail. A version for discussion might be: "A human genetic modification that is intended or may reasonably be expected to alter germ-line cells shall not be undertaken without review and sanction by a body especially constituted to examine not only technical risks but political, social, and moral considerations."

The language covers two kinds of circumstances that may arise: (1) primarily therapeutic use but with a risk that germ-line cells may inadvertently be affected; (2) primarily eugenic use intended to affect the human germ pool and future generations. The first circumstance raises mainly technical issues and calls for reasonable assurance that the risk of affecting germ-line cells is as low as can be assured (given the existing state of knowledge) and is acceptable. The second issue is more than technical because it relates to broader long-term objectives. Unlike the first instance, where a particular patient is waiting for relief, it would be reasonable to require a delay of implementation, even if approval is likely, to give full opportunity for all negative arguments to be presented and considered. Where the intention is to affect the human genome in ways that may prove unalterable, the most cautious and conservative decision-making is called for.

The proposal calls for a special review body that might either provide sanctions for a particular application itself or make recommendations to an elected and responsible public official or group for final decision. The review body might be separately designated for each proposal or might have a continuing role. The exact mechanism adopted deserves most careful consideration because the function being discussed is actually eugenic control. Even to establish a mechanism of this kind is to give legitimacy to the concept and this is itself an issue to be considered carefully.

A third principle might be phrased as follows: "Except as necessary to implement the preceding principles, no restriction shall be placed on research to increase understanding of human heredity and development. To do so might foreclose options that would otherwise be made available to future generations." A principle of this kind is essential if principles like the first two are adopted. In effect, the first two put legitimate constraints on innovation—in the interest of human rights, both individual and collective. The rationale is an extension of the requirement for informed consent as a constraint on human experimentation in general. But if there is a constraint on innovation there must also be a constraint on the constraint, otherwise research and innovation may be seriously curtailed. It is worth noting that new genetic capabilities have come contemporaneously with nuclear weapons and space exploration. Whatever our need for genetic therapy and modification today, it may be much greater if nuclear bombs are unleashed or extraterrestrial colonization is attempted. Our concerns of today should not cripple our capability to achieve the aspirations of tomorrow.

A fourth area that needs early attention is the range of application of any principles that may be adopted. The collective human genome—the gene pool—knows no national boundaries. Whether the objective is to project or to enhance the genome, no effort can be effective that does not command support everywhere in the world. Therefore, "any principles governing human genetic therapy or modification shall be incorporated not only into national law but into international covenants." Moves in this direction that already have begun in the European community should be supported by professionals and other groups in the United States and should be the subject of new initiatives in US foreign policy.

Let me conclude with a response to a question put to participants in this conference by our chairman, Robert Esbjornson. He asked us to consider what is meant by being human—specifically whether it is more human to live by the rule of survival of the fittest than to live by cooperation and caring. A biologist, speaking professionally, can give only a partial answer because the full connotation of being human is not confined to the province of biology. The broader and, hopefully, still expanding connotation of humanity should be viewed as a trajectory, perhaps beginning when our distant ancestors first conceived limited futures and judged them to be good or evil. The full course of the trajectory is probably less certain today than it appeared to be a century ago, largely because we have, ourselves, introduced uncertainty by assuming so large a role in determining the trajectory. However, caring and deciding are as much human activities as blindly struggling for survival, just as dining at Maxim's is neither more nor less human than sucking a nipple. Each is a stage in the human trajectory, retraced in each generation and controlled increasingly by human decisions made in historical cultural matrix. Our expanding options for building upon our biological human quality are an opportunity to continue our trajectory—with a growing necessity and responsibility to increase the God/Satan ratio. This requires us to be increasingly purposeful and increasingly participatory in formulating our purposes and implementing them. Our human thrust, as a community, is to move outward to a wider existence—physically, intellectually, and spiritually. Exactly what this means will not be fully defined except as it is achieved, and then it will only be a platform for a further step toward what lies beyond. For better or for worse, for good or for evil, whether divine or satanic, that is what it means to be human.

Without Laws, Oaths and Revolutions

JUNE GOODFIELD

When I learned I was to give the address at the closing banquet, I was tempted to behave as I have often observed banquet speakers behave; namely, to place before you a well-prepared speech, which I then solemnly read from the podium. But for a variety of reasons I couldn't deal with the occasion in that way. This has something to do with my own Celtic blood, which I'm happy to share, even at a distance, with my dear friend Lew Thomas. It also has something to do with the challenge implicit in the very kind remarks with which I was introduced. But it's also because I was talking with one of your number, who has been so kind in looking after me during these days on your campus, and I said, "What happens at this banquet?" She replied, "Well, you see, it varies a lot, but one thing is certain. We are really all flaked out." I thought, what an interesting evening we're going to have with a flaked-out lecturer facing a flaked-out audience—a situation that clearly called for drastic action. I thought the

This talk was delivered at the closing banquet of the conference and the style of its delivery reflected the occasion. It has been reworked and rewritten to form the final chapter of the conference book.

best thing I could do was to utter those immortal lines from that magnificent film *All About Eve* and wish I could utter them with the lethal force of Bette Davis. The lines are, "Fasten your seat belts. It's going to be a bumpy evening."

For this final particular occasion, however tired we are, I think we have to try to pull ourselves together for one last effort. One of the reasons is that I'm going to come from a direction quite different from that used by any of my colleagues. I am here tonight as someone who is no longer either a professional working scientist (a part of my career that lasted approximately eighteen months) nor a professional philosopher. You can smother me with technicalities, you can stab me with syllogisms, and you can watch me vainly trying to wriggle free. Rather, I come from a direction which I suspect most young people at the back of this room are coming from—nonspecialists from ordinary society. It is primarily to the young that my remarks are going to be directed this evening.

Untutored though I may be in many of these areas, I certainly have a great interest in history and an equal interest, not only in the problems of what it is to be human, but in what actually it is to have, or acquire, moral responsibilities and moral sensibilities. I am going to paint a very broad canvas; I am going to take the widest possible view of the brief I was given—even to the point perhaps of being irrelevant as well as irreverent—a trait you have surely noticed in me from time to time. But I do want to approach our topic from a very wide stance and I want to share with you some of my musings—they cannot be dignified by any other word—about morality. And if I am sometimes appalled, as I often am, by my cheek in presuming to speak about such a profound theme from an unspecialized point of view, I take great comfort in the remark made by the famous philosopher Max Black, who once said in public at a conference I was attending, "If I wanted to know whether an action I was about to do was right or wrong, I wouldn't ask my professional colleagues. I'd ask my wife."

I am interested in pursuing the whole problem of human responsibility, human moralities, and sensibilities, and it's quite useful to try to begin by teasing out some questions: what does this notion embrace; what does it imply; and how does it apply, not only in the professions, but in a wider society too. Indeed how does it—how did it—evolve? Once when I was musing on such matters I asked myself: what is there to be explained about moral responsibility or moral sensitivities? Had I been born in an earlier age, several centuries ago, I would have answered, "Absolutely nothing at all." For it was once supposed that man is a naturally moral animal having an innate tendency towards morality, just as matter had an innate tendency towards certain forms of movement, and that's the end of the discussion. Men might not be divine like God but they were not totally bestial like the animals. They were moral. And so remarkable was this property that Kant was moved to speak of the two things that fill the mind with awe: the starry heavens above and the moral law within.

Today I think we would regard the acquisition of morality not in absolute terms, but as a property emerging out of communities and social groupings. If we look back we will see that moral sensitivity has evolved and changed through time. As we survey the tapestry of human history, we will see, within certain societies and on certain occasions, pools of people with a strong collective sense of moral responsibility. When I once suggested this to a friend and colleague at the College of Human Medicine, Michigan State University, he positively shrieked. "Pools," he expostulated. "You mean droplets." Yet, nevertheless, droplets or pools—certainly not lakes and oceans, I concede—I do have a sense that over periods of time and in certain societies, there has been a gradual evolution of notions and acts that have coalesced, which have had the effect of heightening personal, individual, and collective moral and ethical sensibilities. I find it extremely interesting to ask: how did this come about in the past and how might this evolve in the future?

The evolutionary analogy may possibly be a thoroughly bad one, because so far as I am aware (and if I'm wrong I hope my colleagues in the biological sciences will correct me), animal societies and populations don't regress evolutionally speaking. They do not return to their former state. But so far as morality and ethical sensibility are concerned, societies and humans and professions most certainly do. Through history we have seen the peaks and we have indeed, by God, seen the troughs. Some unkind person once described Texas as a state that moved from barbarism to decadence without an intervening state of civilization. Yet we all know and can point to societies that have moved from barbarism to barbarism *with* an intervening stage of civilization. I find such regression just as puzzling as the progression. It is as if the moment the awareness goes, or the pressures or the conscience or the human response falls away, everything falls apart. So if the price of liberty is eternal vigilance, then I want to argue that the price of moral responsibility is eternal conscience, compassion, and caring and whatever it takes for a sense of moral responsibility to evolve at all, it is going to take the action of those same forces working continuously, if it is to remain.

Please hang on for the moment to these notions of caring and compassion, because I shall return to them. It is clear that if society is to have a moral responsibility (and I apologize for making it sound like the measles), there has to be some kind of prerequisite. In the same kind of way that in the period around 400 B.C. there had to be certain prerequisites for the city states of Ionia to evolve notions of abstract ideas and to set in motion the questions that we still ask in modern science, so too there have to be certain conditions that must be met now. I suspect there has to be a real degree of stability, and certainly a major degree of communal self-confidence. There must be a lack of degradation, whether physical or psychological, though there does not have to be a total absence of physical and psychological pressure. I am not too certain about all this. True, one did see within the communities of the concentration camps, in spite of the terrible degradation, the most magnificent acts of moral and ethical awareness. But one thing, however, is certain: whatever the immediate prerequisites are, there do have to be certain conditions that create higher loyalties than to oneself. There has to be a notion of transcendental values and there has to be not only an awareness of the existence of these values, but a desire to apply and live by them as well.

Now what could all these conditions be? I must make it clear that I am not talking just about those utilitarian features that make a society work, and that when I talk about being morally responsible or morally aware, I don't just mean the awareness of the existence of laws. Nor do I mean merely the recognition of the need for laws in the first place and the necessity to obey them when we have them. For all that adds up to is being aware of the possibility of anarchy. No, I mean something quite different. I mean an attitude that is just one step ahead of the law, or one step beyond the rules of society. I mean an awareness of existing wrongs, things which are found to be plainly intolerable. I mean an awareness of individuals' and society's future needs and a desire to do something about them. Thus, I will ask you young people to make certain that from time to time, you do try to decide just what are the things you find plainly intolerable in the existing human condition and also decide whether you find it possible to go and do something about them. If anyone wants to ask me just what it is that I find intolerable I shall be most happy to tell them later.

Looking again at the gradual evolution in certain societies, at certain times and in certain places, and seeing the coalescing of certain features that led to pools of heightened moral awareness—to those moments in time when a society's sensibilities were deeply engaged—we can easily make a list of those intolerable situations around which people's feelings began to crystallize. We no longer believe in a notion of the divinity of kings, and are prepared to say this is totally intolerable. We have therefore moved to far more egalitarian views. Except in certain places in the world, we have long conceded the iniquity of slavery, a condition predicated on the existence of a divinely ordained ruling class that could have dominion, not only over the earth and the things upon it, but over people too. We find intolerable the existence of child labor; we are confronting intolerable notions of race. And so we abolish these and we abolish them sometimes with horror at the recognition of how we once could so easily have tolerated them. We move along a knife-edge of thought and awareness that is always one step ahead of the law, one step ahead of the existing conditions, one step ahead of society's prevailing concerns.

Matters don't stop there. I argue that if one arm of moral sensibility is an awareness of injustice, another is the desire to do something about it. If we bring this issue down to the terms of our individual lives, in the last analysis moral responsibility becomes a concerned awareness of the likely consequences of our actions—not in regard to ourselves alone but to other people also. Everything that I say here about individuals applies equally to professions and also to societies.

During this conference we heard a large number of parables. I am not going to tell you a parable; I'm going to tell you a story. I worked in Iran for a period of time for a remarkable organization which, alas, has now been disbanded. This was an organization which, amongst other things, was responsible for the paramedical work among the nomadic tribes. At that time, 95 percent of the rural population in Iran had no access at all to medical care, and I should be surprised if the situation has changed very much. In any case, I was standing at the intersection of Ferdowsi Avenue and the main street of Tehran and trying to cross. The traffic was coming at me like the piranha fish in an Amazonian river. The drivers of the cars ignored not only the pedestrians, but also the policemen and the red lights too. And a dear colleague who was with me, observing me dithering on the brink, said, "No, no. You don't understand. You must not look at them." It was my turn to shriek, and I said, "Why mustn't I look at them?" And he replied, "Because then they will know that you know they are there. And they will come and get you."

It seems that when crossing the street in Tehran, the name of the game is moral responsibility and the aim of the game is to shift it. The argument runs like this: if you actually look at the drivers, they know that you must have seen them and it is therefore your responsibility to get out of the way. But if you don't look at them, they don't know whether you have seen them and therefore know they are there. Thus you have neatly shifted the responsibility to them. They have to be careful, so they won't actually mow you down in cold blood. So, for what it is worth, you must cross the road with the supercilious disdain of a camel combined with the insouciance of Sophia Loren. It is incredibly difficult. The point of my story, however, is not my education by the Iranians so much as my argument that within the notion of moral responsibility are the twin pillars of awareness and caring. It is caring to the point of practical action that is, for me, what adds up to the crux of moral responsibility, and it also seems to me that only when the two things come together—the awareness and the practical action—that we find change and get change. So what I want to do in the next small section of this talk is to try to bring all this into the context of first the medical and scientific professions and second, of wider society.

One profession faces outwards to the world—the medical profession. The other faces very much inwards, so much that, for some people, the scientific profession has been a refuge from the harsh realities of the world. Einstein admitted that it was so for him. The morality and moral attitudes of the medical profession were, we believe, enshrined in the Hippocratic oath, and developed by an ethically sensitive group of people to protect their patients. I don't want to be cynical about this, and I certainly don't want you to be cynical, but we know that this belief is strictly mythical. As Jacques Barzun once wrote, the Hippocratic oath has been "in many respects a wonderful exercise in noble futility." If I confidently agree with this, it is because I am aware, as I suspect some of you are, that historically the oath was administered to students not after they graduated but before they were ever admitted to medical school. Its provisions were designed to protect the secrets of the craft rather than the interests of the patients. It defined a set of rules, or of etiquette, rather than of ethics, and it delineated very neatly the guild's relationship to the society.

The oath has come down to us in history in some twenty different versions, from the pagan oath modified for the Christians, through the Arabs, through medieval times into the forms adopted in Montpelier, then Glasgow in the nineteenth century, and then finally in the form of the Nuremburg Code in our century. It was not until the thirteenth century A.D. in the Arabic version that, for the very first time, rules were laid down which said procedures must be solely for the patient's benefit. I would like to know why this provision was first incorporated at that time. I still don't know.

Throughout history there have been no sanctions directly applied to those who transgressed the code or the etiquette. None actually were possible—certainly no legal ones. The only appeal was to a person's religious scruples, or to their sense of honor—whether personal or professional—and at times this was a very strong appeal. The oath was not a set of codified laws. But around the turn of the eighteenth century, a time of which Willard Gaylin was speaking in the course of his address this afternoon, several things happened. A Dutch physician, Boerhaave, became the advocate for the patient and really influenced the direction of the profession, switching us from medical etiquette to medical ethics. This became enshrined in the Montpelier version of the oath when, for the first time, it became administered when one left school rather than before one ever got in.

Various new clauses were inserted at that time, and one is always certain of a laugh in this day and age. It reads: "I will never exact a fee higher than my work deserves and I will always give my services to the needy without fee." Again one must ask why that clause went in at that time and why is it not to be heard now in any society. Whatever the answer to this particular question is, what we were seeing was the evolution of a profession's moral sensitivities. The emphasis came to be less on craft and more on the patients and society, and included a strong obligation on the part of the physician to consider the ultimate consequences of his or her conduct.

All this was occurring at a time when several other things were operating: the profession was becoming more professionalized; society was actually at one of those moments of crystallization; when there was, for some reason that I cannot yet understand, a general increase in ethical and moral sensitivities. This would culminate in many things I mentioned earlier: the abolition of slavery and child labor, the extension of the universal franchise, not to mention the American Revolution. These political rights and attitudes became enshrined in the writings of many people, Kant again included.

I see Kant as someone whose writings encapsulated many things that other people besides him were thinking; many concerns were stirring within his society, though I don't know how aware he was of doing this or even of what was going on. Nor is it obvious that many people at that time read him. I believe, however, that we were at one of those great moments, a point at which someone eventually did say—and in speaking in this way expressed a newly evolving belief—that the way we have to regard people is not to consider them as means to an end but as end in themselves. It was Kant who was to write this and it was Kant who pointed out that we all belong to the great kingdom of ends. As Will Gaylin pointed out this afternoon, this was one beginning of society's awareness of the individual worth of human beings.

So far as the medical profession is concerned, it was this very same period that saw the rise of professional obligation and motivation that now began to enter in very strongly, with an emphasis more on caring than on behaving. Caring becomes the source of both moral obligation and wider action, and extends this obligation from the patient to society too. Thus in the nineteenth century we see evolving as just one hallmark of the medical profession's maturity, their guarantee, not so much of each other's technical competence or the fact that they will share the secrets of the art, but of each other's behavior and morality towards the public. They were, that is, to act as *protectors* of the public interest. The law would hold them to the highest degree of technical care, but now the profession would also hold them to the highest degree of ethical behavior. So it was as the nineteenth century evolved. Thus three protections were built in: towards the patient, towards other members of the profession, and towards society. The standards that now were to be maintained were both technical and moral ones. And this was new in the guild and it was also new in society. Indeed, the medical professions at that time extended this obligation even further. Many of them saw themselves as the prime wedges of change for the good within society.

But things were never quite as simple—and certainly never always as good. The institutionalization of the medical profession in the nineteenth century certainly gradually gave rise to these higher, more transcendental values, carried in on a tide of awareness that was moving in other sections of society too. As the nineteenth century moved into the twentieth, we perceived how the very existence of the professional also began to institutionalize opportunities for professional self-interest. This happened in other professions too. We also see the beginnings of the professional and disciplinary fragmentation that ultimately would lead to a degree of remoteness from society and to a degree of moral bankruptcy.

Now let me turn to the scientific profession, one that evolved into a quite different relationship with society, for a whole variety of more detached reasons. For a start, there were no professional services to be rendered to the public, and its members were committed basically to one common goal not shared by any other profession. This goal was understanding, and there was agreed methodology for arriving at that goal. There was no accountability to society because there was no imperative to account to society. The only underlying ethic was one that subscribed to a methodological commitment—to a means for arriving at understanding, and agreement to state and share, not only the understanding reached, but the detailed processes by which it had been achieved.

In the late twentieth century we see that the confluence of new movements within society has begun to impinge heavily upon the scientific profession. It is, in part, these movements that explain why we are holding a conference here on this particular theme. On the one hand, and externally to the profession, there has been a great increase in political and populist awareness in society. On the other hand and internally to the profession, there has been an enormous increase of our intellectual and technical command of the makeup and processes of the world we inhabit. The merging of this new scientific knowledge with people's new perception of this knowledge—now equated with power—has meant firstly that we are now unable to distinguish as neatly as we had once managed to do pure knowledge from applied knowledge. One such area is human genetic engineering. But secondly, the confluence of these factors has forced the profession to answer, even perhaps to justify to the public at large, just what they do, how they do it, what are the likely consequences, and whether indeed they should do it at all. This was something that the scientific profession had never expected to justify, and now questions of wider moral responsibility begin to impinge so heavily upon the profession, and new ethical imperatives make themselves so strongly felt, that many people have argued that there should be a Hippocratic oath that would hold scientists to certain professional and agreed standards.

By now you are probably aware that not only do I find the Hippocratic oath vague, but I also find it unenforceable. I see no reason why what has been impossible within the medical profession should now be possible for the scientific profession. I have to admit I am also rather suspicious about the value of laws in this case, but not because I think they're unnecessary and should not be obeyed. On the contrary, I think they're absolutely essential and must be obeyed. Laws, however, don't guarantee people's behavior; they only declare that this is a situation that must be taken seriously. Now, since one common tendency is to argue that as it is in the nature of all systems and all professions—whether political or scientific or medical—to be totally self-serving, then surely the only remedy is the radical solution. Have one great big glorious revolution and sweep everything away. I must say I find the oath impractical; I find the law here rather dangerous, and revolution is utterly beyond my competence. Even if this weren't so, I am convinced that moral sensibilities have never come by way of oaths, or laws, or revolutions.

Let me try and tease out the dilemma a little further. There is a widespread assumption, which has a great deal of force, that science is the most appropriate technique for understanding reality. This is an epistemological statement with a degree of totalitarian overtones, but for the moment I'm happily prepared to go along with it. If you put scientific activity in these terms, people often say, "Well, you see, questions of morality can't arise in the operation of science. Questions of morality only really come up in the application of science and so it's nothing to do with us." I think I have met this objection in part by saying that the distinction between basic research and applied research does not apply easily—if at all—in the areas of human genetic engineering. For example, if you are interested in seeing whether a properly functioning gene can replace a defective one in the cells of the bone marrow, there's only one way you'll ever find out—by doing it.

My friend Professor David Jenkins (Professor of Theology at Leeds University), whose opinions greatly influenced me as I prepared this talk, has offered a somewhat more cogent argument. Rather than being a particular class of persons who are finding out things by an absolutely guaranteed and an impeccably purified method, scientists are, whether they like it or not, as much shaped by and shapers of social and human reality as anyone else. And they are now probably at a point at which the older professions once arrived and had to survive; a point at which they are having to consider what to do when a situation arises where human values as society presently determines them clash with traditional professional mores and attitudes.

It is doubly confusing for scientists to have to be faced with this. We are asking them to take value stances in certain aspects of their work, sometimes, Peter Medawar would say, before they even start to do it. And it is confusing because at certain stages and at certain levels there is no way even to achieve understanding as a scientist, except by being as objective, even as amoral, as possible.

There is no problem, of course, so long as the object of interest or investigation is a static rock or an atom or a plant. Nor, as David Jenkins emphasized, must any of us here try to confuse the issue by investing matter or the investigation of matter with moral tendencies. Matter has its own autonomy and that's that. But the problem begins for science, as we have seen throughout this conference, when the object of the scientist's interest has two kinds of autonomy: the autonomy that exists by its being part of a material world, and the autonomy that exists by virtue of its being human. And the clash arises at this point, and not only for scientists, either. We are all involved, and none of us can avoid this clash by saying things like, on the one hand, "Well, in basic science there is no morality," or, on the other by saying, "Will you kindly build morality into your scientific method?" That is sentimental waffle—romanticism—it cannot be done. I agree with David Jenkins; there are always going to be facts that are facts, whether they kill you, whether they enable you to laugh, whether they enable you to cry, to rejoice, or to despair. This is the glory of being human and it is also the tragedy of being human—that there are facts that cannot be mitigated. But one can move on from these facts, and this is what I suppose I want to do next. One cannot evade the fact that the sun is at the center of the universe or that *E does* equal Mc^2 and that this formula carries some terrible implications, or that we now have an enormous capacity to manipulate life. Scientists or society, we have to recognize and face the existence of these new facts. There is no way that these facts can be mitigated. We cannot drive them away by appealing to historical precedent, or parallel situations, or what have you.

Equally, it must never be forgotten that science itself is a human activity in that it is promulgated and maintained by human beings. We have seen that when pools of morality do emergy in history, they arise out of the context of human beings as operatives. Thus, in this day and age, there is no evading the fact: if you are a scientist and if you are interested in human genetic engineering or weapons research, then you have a problem that cannot be evaded: perceptions of value *have* to be faced, questions of morality and ethics *must* be considered, and all have to be reckoned and wrestled with. For those of you who are going into science, I commend to you the advice of Nobel Laureate Sir Peter Medawar in his book *Advice to a Young Scientist*:

> A scientist will normally have contractual obligations to his employer and has always a special and unconditionally binding obligation to the truth.
>
> There is nothing about being a scientist that diminishes his obligation to obey the Official Secrets Act and the Company's rules on not chatting confidingly about manufacturing procedures to bearded strangers in dark glasses. Equally, though, there is nothing about being a scientist that should or need deafen him or close his mind to the entreaties of conscience.
>
> Contractual obligations on the one hand, and the desire to do what is right on the other, can pose genuinely distressing

problems that many scientists have to grapple with. The time to grapple is *before* a moral dilemma arises. If a scientist has reason to believe that a research enterprise cannot but promote the discovery of a nastier or a more expeditious quietus for Mankind, then he must not enter upon it—unless he is in favour of such a course of action. It is hardly possible that a scientist should recognise his abhorrence of such an ambition the first time he stirs the cauldron. If he *does* enter upon morally questionable research and then publicly deplores it, his beatings of the breast will have a hollow and an unconvincing sound.

If you are aware that most of your research is inevitably going to have to be underwritten by the Department of Defense, with its increasing strictures on sharing of information, then you must wrestle with this problem before you begin. Equally, if you are aware that there are a number of questionable things you may have to do in order to get your grants, bringing you into situations that Alan Bullock once described as being forced to operate with a second-order kind of truth, then try, if you can, to see where you are going to take your stand on this before you start. You will have to come to terms with what you are, what you hope to achieve, what your profession demands of you, and what you must do in order to live with yourself. But you must do all this at the outset, because by the time you are deeply enmeshed in science there is much that is inherent in the way it has to be done that commits you irrevocably to a certain course of methodological action.

So what does our present situation call for? I think we need a community of people who are dedicated to doing science as rigorously as possible, but who at the same time are also dedicated to being as human as possible. We also need groups of people both within the profession and society who will help each other bear the consequent tensions. I met such a group of people once and it is relevant to our conference, where our main concern is the direction of biomedicine, that this particular group I met was caught up in what I consider to be the greatest medical triumph of the twentieth century. I'm speaking of the eradication of smallpox, especially as I came to know it in Bangladesh, where a remarkable group of human beings of many nationalities did the science and the technology as rigorously as possible, but did what they had to do as humanly as possible. Over four years they helped each other to share the unbelievable tensions of their human condition and dilemma. As a sideline, you will be interested to know that the three nations that, almost alone, did more than any others to make the eradication of smallpox possible, were the Russians, the Americans, and, how appropriate for this College, the Swedes.

You young people out there, whether in science or out of science, have a number of temptations to avoid. One of them is saying, "Yes, of course, we know that such questions are important, but as we can't answer them, how can we bear them, so don't let's even think about them." I say we have to bear them; I say we have to think about them; I say along with Will Gaylin that we should be thankful there are still some questions that are so appallingly difficult that we have no option except to wrestle with them—and nothing should be done to make these difficulties any easier. But I also say that in order to tackle them what we need is a restoration not only of human concern and of human practical caring, but more importantly of communal support—of rebuilding communal resources for living with these questions. That means developing bridges of a form and quantity between the scientific and medical community and those of us in wider society that will be unprecedented in the history of this profession.

There's no possibility of abrogation here: there's no falling into the temptation of saying, "I'm too busy. It's my job only to heal the sick, make new discoveries, codify new laws, teach a new generation of students, develop new weapons, build new automobiles." This responsibility cannot be shifted from one group to another, for this is acting as the Tehran taxi drivers do. It is to shift the moral burden onto others.

Besides, there is a second temptation: believing that the mere enunciation of the problem is enough to discharge one's moral responsibility. I recently attended a meeting at Basel, where a group of young (and not so young), concerned, affluent Swiss were attacking the drug companies for not having eradicated leprosy. They said if all the drugs are given away free, leprosy will then go away. I could tell them that this alone is by no means enough: the smallpox vaccine was mostly donated to the eradication campaign and the bifurcated needle, made available at well below cost price. Still, what is deeply needed over and above such gestures, is personal commitment to act, both on the part of the societies in the developing world and in ours too.

The important point of this particular story is that the group bashed away for a good seven hours at their "enemies," who sat there and took it. And then, finally, someone said to this group, "Okay, well all right, will you please tell us what we should do?" And the answer that came back was, "Oh, that's not our job. We are only here to criticize."

It is simply not good enough to attempt to discharge your moral responsibility in this facile manner; to rest comfortably in the illusion that criticism is the proper extent of commitment. Moral responsibility means more than arguing about these matters, either in terms of an academic course or from the sidelines of society. Nor is it enough to say, "I am disturbed; I'm bothered; I need a dose of ethics." One must go and do something. The extent to which courses in ethics can enhance moral sensitivities has always been questionable. Perhaps by now it's no longer questionable. We know courses may help but without action they achieve little.

I hope by now you see why I have felt that one could never isolate the questions we have been discussing in the last two days from the total context of the society in which the professions are operating, nor from the totality of the human dilemmas we all share. Thus, though this talk may stand in complete contrast to those of my distinguished colleagues by its generality rather than precise detail, I do not want you to forget the continuous and mutual interactions of the scientific and medical profession with society. We create, mold, and determine each other to a great extent. So if we feel that earlier moral attitudes are either missing from the medical profession, or not yet incorporated into the scientific, we must also ask ourselves, as we did with all other issues of the rise and fall of moral sensibility, what is the totality of forces that is bringing this about? How far are we both individually and collectively responsible?

We all have to ask value questions, to see where they arise, and we all have to try to establish where they are coming from and why. Why do we have to do this? Because these value questions alone provide the human challenge and the human sustenance. Yet the prospect of having to do this disturbs us, and again one must ask, why? It's easy to understand why it sometimes disturbs scientists, because as a professional group they've never really had to do this before. They are fearful that it might disturb their methodology; they also feel uncomfortable dealing with uncertain questions, untidy answers, and most especially, with problems that do not appear amenable to permanent solution. The very notion of a permanently insoluble problem is an anathema. Yet it is a sign of the beginning of wisdom and of human maturity, to recognize that there are indeed certain problems that will recur in new forms over and over again. To this extent there are never likely to be any permanent answers, only permanent problems. Certainly we cannot tackle these problems by throwing technology at them.

We in society have also felt somewhat unhappy in having to deal with such questions. Perhaps materially we're far too comfortable and it's easy to just let things slide. Perhaps—but once again I'm just musing—there is a crucial difference between the ambience of society in the past and now, one that is fundamental. In past times we were so certain of the existence of a future world that we could happily tolerate the existence of uncertainties in the present. We are now so uncertain of the existence of a future world that perhaps we grab at any kind of certainty we can. We hang on to it because psychologically we need to do so and it's all we've got. I suppose in a way I'm saying what Lew Thomas has said more beautifully with his inimitable and poignant use of metaphor. Please read the last essay in his latest book, *Late Night Thoughts on Listening to Mahler's Ninth Symphony.*

Yet, on the other hand, we must also recognize that disturbance is creative. Scientists may not like insoluble problems, but most of the great discoveries in science started with just that—apparently imsoluble problems. Disturbance has been a creative force in society. Disturbance has probably been, as Kenneth Clark showed in his series "Civilization," the most creative force in civilization. I now argue, therefore, that to be moral and to be human is to welcome disturbance—to be prepared and ready to wrestle as hard as you can, with whatever resources you have, for as long as need be, with the dilemmas you find; and follow your feelings and your troubles and excitements and your pain about being human. And once again I'm merely, I suppose, reiterating what Will Gaylin said: Some questions have to be agonizing and it is the privilege of being human that we have the capacity to recognize this and deal with this.

The main theme of our conference was once stated by George Steiner in his Bronowski Lecture, when he spoke of the new discoveries in science as laying down a whole series of moral ambushes for man. There, he was really restating the old and perpetual problem of the myth of the Fall, the myth of Prometheus, and how it is that getting new powers always seems to lead us, if not to disaster, at least to terrible problems. This is the constant human dilemma, and being reminded of it again today, on this occasion, with this particular topic, in this particular setting, we are being forced to rethink, not only about what it is to be human, but what sort of society we are creating and what sort of humans we want to be.

Socrates, as always, had a word for the state we are in: it is *aporia*. It is a state of not knowing one's way, not knowing where to place one's feet. And it is the essence of human and moral advance, as David Jenkins emphasized to me, that people are sustained through *aporia* and do not get frightened by it and do not simply fall back upon old precedents that have limited force in new historical situations. With our new science, we are in a totally new historical and scientific situation and we cannot revive the example of Galileo and the Church, or Lysenko and the Russians, or Victorian professionalism, or Marxism, or the Ten Commandments, to help us. None of them alone will do, although they may all be drawn upon. This is particularly difficult for those people who interpreted the scientific life in a narrow way and it's painful for those of us who didn't expect science to get us to the point where we would be forced to think about these things.

But why should we, in the late twentieth century, think we should be exempt from these considerations? It happened before in human history at the time of Copernicus, in the time of Darwin, and if we believe—as I do—what Nobel Laureate David Baltimore is saying, it is inevitably going to happen to us now with the problems posed by bio-ethics and biomedicine. For, he said, these are, in certain respects, even more critical than the problem of nuclear power. Certainly nuclear power forces us to contemplate a wholesale destruction of the world; but with the new biological powers we are forced to contemplate the possibility of a new kind of human.

There is now no way to avoid the requirement of considering what it is to be human. Of course we could just say, "Oh no, no. Being human is actually just being very very dull and we must be careful to keep it this way, not so we won't get excited—though that's bad for our blood pressure—but so we won't suffer. Let's just keep it all calm; stay where we are, and not even get worried about such problems."

If that is our attitude, so be it. We keep our nice comfortable existence with our very dull souls but we also, of course, are going to risk heaven and we are going to risk hell. Actually, we're probably going to miss making any kind of progress at all, for progress comes as a result of human aspiration, vision, ecstasy and the personal suffering that attends the creative process.

Finally, I want to swing back to those pools; those pools or droplets of human moral responsibility and the events that caused people and circumstances to crystallize together. I suspect we might now find ourselves in a similar moment. It is just possibly a point in time where things might begin to coalesce once more because circumstances, whether thermonuclear destruction or our capacity to manipulate life, are forcing us to think about the world and our place in it. If this is indeed one such moment, then I say we are mightily privileged to be alive and to be participants at this stage of the human drama in which issues, whether they be economic or political or scientific, are held right up in front of our very faces, so we are forced to reassess ourselves and our values.

Right now the major pressure is coming from within science—brought about by our capacity to manipulate life, which sadly is more than equalled by our capacity to annihilate it. And this of course poses tensions and cultural contradictions of the most stupendous proportions. But the pressure also comes from within ourselves and from what we outside the professions have allowed our society to become. If there has been one message I've tried to keep running clear and cool throughout this conference, it is that whenever we begin to talk about these things, we must never make the fundamental mistake of trying to extrapolate the consequences of the work and the activity of one single group of human beings out of the totality of society. If we are where we are, it is because *we* have chosen to be there, not because scientists alone have taken us there.

By now you are probably saying, "Oh, my God, she's uttered so many generalizations as to be utterly and totally useless." So where do we go from here? Well, I think one direction we take from here—at least this is my small direction from here—is to start facing outwards a little bit; to refuse to accept the fragmentation of political and intellectual and professional life that, with a smile, turns back continuously upon itself. We are being forced, I think, to shake ourselves out of this stance and look at not only other people next door and those down the block, but those across state lines and over the ocean. Otherwise the incestuousness, the introvertedness, leads not only to a total bankruptcy of ideas and a chilling of the spirit, but to sin, as Martin Luther defined it. "*Cor incurvatus in se*" was his definition: "the heart turned upon itself." And in this celebratory year of his we do well to ponder this.

This definition of sin holds equally for the individual, the profession, the society, or the political institution. We have to face outwards to where the problems are. You will find that where the problems are is quite different for each one of us. You have to reach out to those things in the human condition that you find intolerable and then you have to be prepared to do something about it, whether your concern is leprosy all over the world or nuclear warfare or the soup lines which, I understand, now exist once more in American cities. To do this—to be prepared to do this—is to have the courage to be human, to find out one's values and act upon them. This, of course, involves a great act of moral imagination, but at the same time it is also what life and science is all about. As Jacob Bronowski said, "It is not the business of science to inherit the earth, but to inherit the moral imagination, for without it there will be no earth, no men, no women, and no science."

So this is an invitation to be human in a rather different kind of way, I think, from the ways we have been talking about during these last two days. It is an invitation to have and show feelings; for it is an invitation to believe that there is going to be a future; an invitation to make certain that there *will* be a future and that your children and your children's children will be there to inherit it. To be human is to embrace the freedom that you have to do this and the challenge to do this and the privilege of doing this right from the start—the magnificent start—provided by such a peaceful place as Gustavus Adolphus College. If we don't want to move this way, we don't have to, but we've got to accept the collective consequences, one of which is that we may all, one way or another, perish. If we do decide to embrace the vision and incorporate into our individual and professional lives those elements of heightened moral awareness with practical action, I have straightaway to warn you that there is no guarantee of success whatever. There is only a guarantee of a great deal of personal cost, much personal disturbance, and inevitably some suffering. One parallel, as I've said, is human creativity, whether in science or in art. Another is in loving—in falling in love.

With one exception, all my colleagues have quoted from the Bible, and as befits a Welsh minister's daughter I have no intention of being left out. Unlike my colleagues, however, I am going to quote from the New Testament and I am not going to reveal in full the trouble I have had to get the St. James version in front of me today, on this campus. I am sure most of you know the piece I am about to quote, from the Epistle of Paul to the Corinthians, chapter 13.

Though I speak with the tongues of men and of angels, and have not charity, I am become as sounding brass or a tinkling cymbal.

Though I have the gift of prophecy, and understand all mysteries, and all knowledge; and though I have all faith, so that I could remove mountains, and have not charity, I am nothing.

Appendix: Conversations at Nobel XIX "The Manipulation of Life"

Following the Lecture of Lewis Thomas

GAYLIN: At one time, access to a physician was as likely to contribute to your death as to prevent it, so to buy the services of a physician was not necessarily a privilege. In the last fifty years medicine has become a life-saving profession, not just a matter of giving comfort. Dr. Thomas, would you address the question of access to medical care? How are we to distribute fairly the fruits of our progress within the entrepreneurial system we now have?

THOMAS: The entire profession, including the nursing profession, is worrying and debating the question of how to provide access to high-quality health care in life or death situations for everyone. It is now clear that simply to train up more physicians and have more medical schools is not going to work, if we keep on doing it. People who study the medical profession from a social science perspective have concluded that we must be wasting an awful lot of our time, when about 80 percent of visits to doctor's office or out-patient clinic are for either illnesses that would probably go away within the next twenty-four to forty-eight hours, or are emotional distress;

This section records comments by the speakers in response to one another and to questions from the audience.

anxieties about being ill rather than real illnesses. Perhaps half of the other 20 percent are illnesses for which the technology of medicine is absolutely essential to prevent incapacitation or death. How to devise a system to prevent the 80 percent from crowding out all who really need quality care is an open question.

We have large segments of the population in this country who never see a doctor, such as people in the ghetto areas in cities. There isn't any equity in the distribution of care.

Perhaps state or federal government agencies should underwrite a substantial part of the high cost of medical education in return for a period of four or five years national medical service in places where doctors are most needed.

I worry still more about the health problems in the Third World, where disease problems are critical. The only solution to the population and overpopulation problem in the poorer nations is to do something to provide adequate health care for newborn babies and children under the age of five, 70 percent of whom die in some societies.

What this country needs is more nurses of either sex, highly trained to cope with most of the problems that now confront doctors, or to refer patients quickly to more specialized care.

If we had an international health service in which both European and American young people could serve a period of time after medical school in impoverished countries, I would like to see the nursing profession involved at a high level in that service.

GROBSTEIN: Since medical disciplines tend to separate holistic and reductionist approaches to medical problems and few people have found a way to integrate them, what advice would you give to young people about their preparation to function as health care professionals?

THOMAS: All I know from my own experience is that when

I was young I knew that medicine would ever remain as it was in my father's practice. It has changed and what it is like now will also change rapidly and will be a totally different profession within the next five to ten years.

LEBACQZ: Why are you so pessimistic about the bureaucracy related to war and yet so optimistic about the bureaucracy related to medical research and practice?

THOMAS: There is in the world today a network of working scientists who, though they receive funds from their own governments, work in a totally nonnationalistic way. The biological scientists are in close touch with each other across national boundaries, except that of the Soviet Union. Young molecular biologists at my own institution seem to know day to day what is going on in Melbourne, Edinburgh, London, or Paris and work with their colleagues in foreign countries as if they were down the corridor from them. I trust in the persistence of this because I see no other way of having science done. Science is entirely a communal effort. I hope that governments don't catch on to the fact that they have on their hands an international network that is genuinely subversive in regard to strictly nationalist concerns. I wish we had some way of drawing in the Soviet biological scientists. They would profit from closer contact with their nonnationalistic counterparts in this and other countries.

THOMAS: One of the questions I received raised the issue of the artificial heart. How can we possibly afford to go on with that development? The obvious answer is that we can't. I see no way in our kind of society we can make our way around not just the financial problems, but the ethical problems involved in access to it. The existence of the artificial heart means that we need more research and we need it in a hurry. The prospects for that research are exceedingly good. If we keep on doing the kind of basic research that I've been urging, we should be able to obviate the need for an artificial heart. An array of drugs is beginning to emerge that are specifically antiviral in their properties and do not seem to be producing the side effects that one might expect from drugs that can act at that intimate level of cell life. I have the highest hopes for the continuing development of drugs directed against viruses that will be as effective and easy and safe to use as the antibiotics that we've been using for the last half-century or so against bacterial pathogens.

Coronary heart disease is another form that will require the choice between clumsy devices such as the highly expensive artificial heart and something very much cheaper that will enable us to prevent or reverse the process. As the result of basic research done in more recent years, it begins to look as though the circulating particles in the blood are at least partly responsible for blood coagulation and are intimately involved in the kind of damage to the coronary arterial wall that leads to occlusion. It has been discovered that platelets do damage by becoming sticky, and that it is possible to reverse that stickiness by interrupting the chain of events that leads to the synthesis of prostaglandin. There are classes of drugs that look as though they will turn out to be highly effective in preventing the stickiness of platelets. We may have our hands on a class of drugs that, when used in the middle years by susceptible people, will have the effect of preventing coronary heart disease. We will therefore not be required to do cardiac transplants or insert artificial hearts.

I think the same thing can be said for cancer, which now costs society an immense amount of money for the surgical-radiological and long-term hospital treatment. I would guess that as a result of research that's going on now at the benches of basic biological scientists, we are beginning to learn enough about the intimate details of cancer so that the genes that are probably switched on, which are responsible for the transformation of normal cells to cancer cells, are now recognizable.

There are examples of advances in controlling once

virulent diseases. None of my colleagues in any of the departments of medicine in New York have seen a case of tertiary syphilis for the last fifteen or twenty years, and certainly not syphilis of the brain. Rheumatic heart disease has almost vanished as a public health problem in this country. That happened as a result of antibiotics which themselves happened as a result of basic research. There is something to look forward to as we continue to do good solid basic research in the biomedical sciences. And one of the things to look forward to is that the treatment and the management of diseases is bound to get cheaper rather than more expensive.

Following the Lecture of Karen Lebacqz

GAYLIN: Dr. Lebacqz, you were saying we are using the wrong paradigms and asking the wrong questions, such as those about human nature, personal autonomy, human dignity, the distinctions between rights and responsibilities, the obligations of individuals versus the responsibilities of society, the nature of integrity, and the question whether or not the human species is mutable. If, in relation to the development Dr. Thomas described in his lecture this morning, these are not the right questions, what are some of the right questions? And what are some of the proper paradigms?

LEBACQZ: It is an important question. Let me tell you a story. A fisherperson went out day after day with a net. Fishing nets, as you know, are made with string or twine woven so there are holes in the net so that it becomes flexible, and usually the holes are all of the same size. So this fisherperson went out day after day and cast this net into the water and at the end of the day brought home a catch of fish. After doing this for years and years and years, eventually she or he concluded that the only fish in the sea were the ones that got caught in the net. The idea that there might have been fish that were too small or actually managed to swim right through the holes in the net was considered at one point, but eventually dismissed, and the idea that there might have been fish so big that as the net was thrown they simply swam under or around it was also considered at one point, but eventually lost. and so after a time this fisherperson convinced everyone that there were only fish of a certain size in the sea.

That's my answer.

I am not arguing against the use of logic and rationality; I am only suggesting that as we use them we may be missing other paradigms that might be available, which come to us in a form we do not readily recognize because we, being Western, middle-class and highly educated in the ways of Plato, Aristotle, and Kant, do not easily perceive the truth in those other forms.

ESBJORNSON: Are you suggesting that from other cultures there may be more intuitive, imaginative, and playful kinds of thinking we have not fostered except in the arts? Do we need new paradigms in ethics and related fields so we will take such modes into account?

LEBACQZ: Yes, I am. As an ethicist trained in the Western traditions, I don't know how to use new paradigms adequately. I hope we might learn how.

GROBSTEIN: I look at this from a public policy point of view. In our political system we more often establish policy by accommodating rather than by arguing about paradigms. People come together in support of a way of dealing with a contentious issue and do not argue their rationales very long. What convergence would be achieved by the approach you are suggesting? What groups within or beyond our own political system are likely to converge?

LEBACQZ: There are not many countries in which we would hold a conference on the issues we are talking about, which are relatively elitist questions. If we were to take seriously the voices of the oppressed, including those right here in our own society, most of them would come before us with cries of pain, saying, "These are not the issues for us." Their voices might be important for us to hear. The issues are important and we should work on them, but we need also to be challenged by the voices and stories that don't seem rational to us.

GROBSTEIN: Of course it is only this kind of society that produces this kind of problem through its scientific activity, and it is this kind of society that should assume responsibility for finding appropriate solutions.

LEBACQZ: It is possible that we make a mistake when we think of our society as an independent entity rather than as related to the entire world and dependent on it. Our paradigms need to take that international context into account. Maybe some of these problems are really peculiar to a small portion of the world and maybe we would not be confronting them if we hadn't made decisions along the way that are based on our Western paradigms.

GOODFIELD: I am tempted to resolve this issue by throwing the word "paradigm" out of this building. Aren't you talking about values? Why are we so incredibly embarrassed to talk about values? Scientists find it easy to handle questions of what we know and what we can do with what we know but, like people outside of science, are extremely embarrassed talking about values.

I remember one particular occasion in England when we were discussing some issue about putting some new health organization in a deprived area. One of the doctors, bless his heart, got up and said, "Look, before we get on with this, shouldn't we really talk about what values we are trying to achieve?" It was as if we had told a dirty story. The chairman said, "Oh, that is fine, that's all right, but let's get on to the next item, which is finance."

LEBACQZ: I am aware of the problem of using "paradigm," but I decided to use it precisely because it has entered the public arena as a shorthand way of talking about deeply held modes of thought that structure our ways of thinking about values. I don't think the term "values" is a substitute for paradigms, although I do agree that we need a lot more value discussion.

GAYLIN: Dr. Lebacqz, how would you jolt people into a healthy skepticism about the scientific method?

LEBACQZ: I would encourage a radical depth of skepticism, not only about particular methods, but about the hypotheses we are trying to verify with those methods and even about the entire way we have undertaken the endeavor.

ESBJRONSON: I saw a television program about a psychiatrist treating Mexicans who had settled in Los Angeles. He got nowhere by treating them with sophisticated methods, so he went into the Mexican mountain villages and learned something about the paradigms or basic structures or mental sets there and then was able to help them. Are shamans, the "witch doctors" of such primitive cultures, able to teach modern scientific doctors anything?

GAYLIN: Among my psychiatric colleagues—we do not claim to be a scientific profession—the rule is that nothing succeeds like success. If dancing around my patients nude would cure their schizophrenia, I'm ready to take off my clothes.

There is a dangerous tendency among the intellectual elite to romanticize primitive medicine. People in Africa do not do so. They want penicillin for their children. I would caution against romanticizing practices, prayer, and meditation. There may be something in all of them, but we have been terribly remiss in supplying the very good technologial aspects of modern medicine to the impoverished.

THOMAS: You are worried about where the elitist scientists may be taking us without enough of us being consulted about the direction we are being taken. Isn't this really an anxiety about technology? The term genetic engineering implies that we are talking about technology, which is what you do after you have understood the underlying mechanisms of something in nature. I don't think there is any level of knowledge about how nature works that we cannot handle and should not seek. It is an urgent matter for us to understand as much as we can about the human genome and how it works, not because we are going to use it or sell it, but so we can gain some understanding of where we are and begin fathoming what I view as the utter strangeness of nature. If we keep at it we may reach a level of comprehension that will at least allow us to grow up enough so that we can stop killing each other. I see the hazard of an antiscientific intellectual movement getting off the ground if we mix up technology and science.

LEBACQZ: My concern is not about where science may be taking us, or about its consequences. I am concerned about how we think about the enterprise as we go into it. And in that connection I question the common paradigm that distinguishes between pure and applied science, between science and technology, as you put it. I am arguing that there are lots of things that we should know that our very scientific paradigms are making us *not* know; they are making us closed to some forms of knowledge to which we need to be open.

GOODFIELD: Dr. Thomas is making too easy a distinction between basic science and technology. In many areas, especially in biomedical science applied to human subjects, this distinction breaks down as we seek knowledge.

ESBJORNSON: Dr. Lebacqz, you received several questions from the audience. First, why do you think we should limit knowledge?

LEBACQZ: I don't; I think we should expand knowledge and honor forms of knowledge that we do not always honor in this culture.

ESBJORNSON: Are we too weak to make any difference in creation and is that why God rebuked Job?

LEBACQZ: Nonsense! We are not too weak, we make a good deal of difference in creation. Everything we do affects the world around us and affects who we are. We are not being rebuked for being weak or for making or not making differences.

ESBJORNSON: Does the Bible say that God is female?

LEBACQZ: The Bible is a record of human communities doing their best to put down, for future generations, what they understand to be their encounter with God. In that record, God is imaged as male, God is imaged as female, and God is also imaged as whirlwinds and other things that, so far as I know, have no sexuality.

ESBJORNSON: What's wrong with genetic engineering?

LEBACQZ: I didn't say there was anything wrong with it. I am concerned about how we think about it, not whether we do it or not.

ESBJORNSON: Is there accepted knowledge that we can use?

LEBACQZ: Yes, of course, we can and should make use of all the rational, logical, accepted knowledge that we have. We simply need at the same time to exercise some suspicion about the limits of that knowledge, and not to forget to be open to other forms of knowledge or revelation.

ESBJORNSON: If we try then to be open to that revelation, how do we distinguish new revelation from emotional babble?

LEBACQZ: An excellent question, to which I do not know the answer. I don't know how we distinguish what is a true revelation from what is simply an emotional outcry. True revelation probably comes from the oppressed.

ESBJORNSON: Can especially the righteous people ask the wrong questions?

LEBACQZ: Yes, especially those of us who think we are righteous, myself included.

ESBJORNSON: How do I identify the right questions?

LEBACQZ: Again, perhaps by listening to the oppressed and the questions they would raise to us.

ESBJORNSON: Does the shift from the language of justice to the language of creation mean that Job's question about justice was the wrong question or does it simply mean that there is an unexpected answer to the right questions?

LEBACQZ: I think it means probably the latter: an unexpected answer to a right question.

ESBJORNSON: What is the new paradigm that you are offering if you don't like the rational, logical ones that we have had?

LEBACQZ: It is my parabolic paradigm, in which the challenge or the openness of the story leads us into a kind of wonder and reverence. Someone asked me, "Why don't we simply ask God for the answers?" I think that's a wonderful idea, and my only caution there is that we have to be a little careful about knowing what God's answer is. So I will tell you one more story, the story of the man facing the flood. As the water came up to his window someone came by in a canoe and said, "Hop in, I'll take you to safety," and he said, "No, no, no, I'm praying for help." And then as the water got up to the second story and he's hanging out the window, someone comes by in a rowboat and offers to take him to dry land and he says, "No, no, no, I'm praying to God for help." And finally the water has crept all the way up and he's on top of his roof and on tiptoe and barely breathing above the water and a rescue helicopter comes by and drops a line and he says, "No, no, no, I'm praying to God for help," and, of course, he drowns. When he gets to heaven and confronts God he's very angry, and he says, "God, why didn't you help me when I prayed so hard for help?" And God says, "Hey, I gave you three chances. What more do you want?"

GOODFIELD: Dr. Lebacqz, might the monster in the closet be those circumstances that force both parents of a family to work in order to cater to the needs of their family and to surrender in those critical preschool years the opportunity to imprint their offspring with the vaules and priorities they cherish? I suspect that the monster in the closet is just this. There are only two times in your life when you are quite irrationally crazy about yourself and these are vitally crucial times: when someone falls in love with you and when you are a child. Sustenance at both times should, if possible, be in no way curtailed. The crucial years of children's nurturing and care are the preschool years and I am extremely unhappy about any circumstance that deprives children of such care.

Following the Lecture of Christian Anfinsen

THOMAS: Dr. Anfinsen quoted me as having said something overly optimistic about the prospect for medicine leading to what I carefully said was a relatively disease-free society; and he expressed the concern that this might do us more harm than good because of the overpopulation problems that would develop. If we got rid of that class of disease we regard as the important ones, it would have only a marginal effect—perhaps add three or four years to the life span. And it wouldn't take care of the 56,000 or so deaths that occur on the highways every year or result from alcoholism, homicides, and suicides.

If we did something to learn about the diseases that afflict the huge population in what we call the Third World, and to introduce technologies to prevent infant and early childhood deaths, the net effect in a few years would be a leveling off of population growth. Most of the pressure toward growth is not so much to have five or six strong sons as to have one child out of seven survive the first five years. Until something is done about the apprehension that children will be lost early in life there will be little incentive for undertaking birth control measures.

What would happen if we also introduced an effective pharmacology to get rid of long-range parasitic ailments (schistosomiasis, filariasis, trypanosomiasis) that don't kill people early but incapacitate and weaken them? We would do a lot for productivity and for human happiness. If we would extend the kind of medical care we have here to the Third World and export a lot of bright young people trained in primary medical care, we could do a lot of good without a major addition to population growth.

I agree wholeheartedly with Dr. Anfinsen that we should be focusing our attention on application of genetics in agriculture rather than on the problems of making the megamouse or the human with long arms. If we could do something about malnutrition in the Third World we would be yards along in solving the population problem.

GAYLIN: I would like to pierce Dr. Anfinsen's gloom; I hate to see such a lovely man so pessimistic. Is your concern about recombinant DNA simply an awe in the face of something new and powerful or is there something special about genetic material that makes its potentiality for evil so extraordinary that it depresses you?

ANFINSEN: What is special is that it is people that are affected. I am convinced that the change in the human genes can be brought about experimentally and become more and more possible as time passes. I am concerned only about changing the human genome and thus affecting evolution. I can see introducing a particular gene into a whole series of women who would contain the gene, so that after a while we would have a subspecies and then, by introducing still other genes, get further changes.

GAYLIN: With successful gene implantation, not only one person at a time would be affected, but the offspring as well, and we would not need very many to set up a population. What you visualize is making a mistake by inserting an awkward gene next to the one you intended to splice.

ANFINSEN: This is why I predicted that doing this was unlikely to happen at any time in the foreseeable future. This is "world-class" research that would require the expenditure of a very substantial amount of money by governments, not something "crazy scientists" are going to do in their basements. I doubt that we will invest in it.

ESBJORNSON: Dr. Anfinsen, elaborate on ways of using genetics for controlling population and increasing food production.

ANFINSEN: Increasing food production is easier. It is a matter, for example, of implanting genes or splicing genes into bacteria that populate the roots of plants and produce ammonia from atmospheric nitrogen, thereby producing a healthy plant that doesn't need fertilizer. How to use recombinant DNA methods to affect population is not as clear.

I cannot buy Dr. Thomas's view that getting rid of tropical disease and producing more food by advanced agricultural methods would not cause a large population increase. If the infant death rate were cut down in places like Egypt, where only one in four survives, the world population would increase considerably. We would reach a new plateau where there wouldn't be enough food to go around. Perhaps we could give every family that maintains a 1.8 child average some sort of prize, like a new Buick.

GROBSTEIN: We realize that to increase the investment in agricultural research to the level of what we now invest in medical research would require a very sizable increment, or a transfer of funds from one program to the other.

We have no really good mechanism for making such decisions. In the United States we do not often raise the question of the relative investment made to agricultural, health, and military science. In Congress these budgets are considered separately by different committees. I presume you aren't suggesting that scientists should be either persuaded or told to shift their research from health to agriculture.

ESBJORNSON: Will the effect of patenting recombinant DNA products affect the flow of scientific information?

Will the increased ties between bio-industry and scientists mean that genetic research will be governed by the profit motive? Will sharing of information be diminished by commercial competition?

ANFINSEN: I can say something about this from what I have seen at Johns Hopkins University in the past year and a half. Industrial corporations are really uptight about projects that involve organisms, which can so easily be transferred. They don't want us to transfer them without getting a signed statement that they will be used only for academic purposes. I don't think it will change the relationships between industry and universities very much, because industries want to support research that will make them money, and universities want money to support the research they want to do. There is a trade-off, and it is not very evil.

ESBJORNSON: A member of the audience asks, "Does international competition, whether military, economic, or nationalistic, present the greatest threat of genetic control getting out of hand?"

ANFINSEN: I think it very, very unlikely. In the area of biology and DNA recombinant techniques and genetic engineering, it's very difficult to think of a situation in which knowledge and application of some genetic idea in one country would frighten another country sufficiently to make it feel that it should compete. These things come slowly, and people who are in the field of science in general read the same journals. A piece of science doesn't exist until it's published. There's no sense doing an experiment and putting it in your drawer, becuase if you want to become famous, you have to publish it. Since it is published and these journals are circulated throughout the world, a scientist discovering something interesting in genetics in Princeton, New Jersey, will immediately be joined by people in Istanbul and Moscow who, very shortly thereafter, will be doing the same thing.

Following the Lecture of Willard Gaylin

LEBACQZ: In Genesis *Adam* means "humankind," not an individual male. It is possible that God is imaged as both male and female here, and that the creation of humanity in the image of God is a reference to sexuality, not to our creative abilities.

GAYLIN: I am more than happy to accept the androgyny of God, whom I, unlike Karen, do not visualize in any image. I found among my colleagues at Union Theological Seminary a unanimity of opinion that the Genesis story deals with rationality, reason, and autonomy.

LEBACQZ: Part of the issue for me is, in what should we have faith? For people who want a nonanthropocentric view, James Gustafson's *Ethics from a Theocentric Perspective* provides the view that humans are not the measure of all things and that all of those wonders of nature are wonderful in themselves. In God's view, whether humans exist or whether we are wiped out by a nuclear holocaust and we are gone, does not mean God is gone.

ESBJORNSON: In questions from the audience there is expressed some uneasiness about exclusive dependence on Judeo-Christian documents as cultural sources. In this time of cultural dialogue and interchange, should not the tradition of the other two-thirds of the world be given at least some weight in considering the future of the human race?

GAYLIN: I certainly agree that the more educated and sophisticated and open we are the happier we would all be. I do find that when I turn to other traditions, I am less comfortable, simply because I am such an amateur in knowledge of those traditions. Therefore, I speak from my tradition, although I am open-minded and would like to have the knowledge of other fields that I do not have.

ESBJORNSON: I have often wondered whether we humans

are the last word in creation. Do we fear being second best to another biological being? What if we become aware of the existence of forms of life in outer space superior to humans, or what if we create our own successors? Is there a fear among us of something better than we are?

GAYLIN: Suppose we could make a robot capable of thinking, creating, talking, loving—in *every* way like a human being. If the robot were indeed imaginative, creative, poetic, and capable of all the traits I previously mentioned and was able to recreate its own kind, I would invite him to dinner. I would welcome such a robot into the family of the humans, for in creating such a "robot," we would have created a human being.

ESBJORNSON: Why does the desire for human modesty and caution and awareness of human capacity for evil negate a sense of awe and appreciation for human creativity? These coexist in us. Are you comfortable with the coexistence?

GAYLIN: Totally. I am only uncomfortable when there is *no* ambiguity and *no* ambivalence. I distrust simple answers to complex philosophical problems.

ESBJORNSON: There is a cluster of questions from the audience on the anthropocentric emphasis on human dignity. Do you realize that man could not have his world as he sees it without the other animals and trees? Is it not by comparison with animals and plants that man gains his sense of dignity? Shouldn't special dignity be ascribed to them also?

GAYLIN: I agree. It is still essential with an anthropocentric position to respect and treat with gentleness all these things we value in our environment. I was trying to make a more subtle and sophisticated point, that although the value of these things must not be minimized, they primarily serve the perceptions, pleasures, and purposes of human beings. We would be diminishing ourselves should we make ourselves unkind and arrogant. I was addressing the problem of self-hatred which as an analyst, I see as a greater danger than pride.

THOMAS: I am not sure we should think that animals have no minds similar to ours. I have known three dogs almost incapacitated by guilt!

GOODFIELD: I am hearing such optimistic assessments from you all that the only thing I can do at the moment is say, "Well, with all this optimism, why are we here? Is there no problem?"

GAYLIN: I do think there is a problem. I am distressed when an authentic area of anxiety is displaced to a non-existent myth about creating a "demi-life" that is going to empty the garbage for us. I do not see genetic engineering as a danger. I see atomic holocaust as a danger. I want all the displaced energy available to deal with the *real* problems caused by the rather mundane low technology of the past.

GOODFIELD: I do not have the widely shared fears about genetic engineering, but I do have a lot of worries about the human condition.

LEBACQZ: Where do we locate what is distinctive about human beings, or what might be considered to be our dignity? For example, we find in the prophets and the New Testament a different understanding of the dignity of humankind—namely, choosing how to use the power we have. You have drawn on biology and psychology. Are there also other standards for choosing where we go for our understanding of human nature?

GAYLIN: I have enough faith in the dignity and specialness of our species that I would predict that from whatever professional perspective you start you would find lists of traits that make us different. As one trained in biology and psychology I draw on those roots. If I were a geneticist or poet I would draw from distinctive traits seen by them. I have enough confidence in the specialness of our species that I think that wherever you start you would end up singing a hosanna for homo sapiens.

ESBJORNSON: Would you please characterize the nature of the complete person?

GAYLIN: No, I will not. We have this capacity to be what the nineteenth century called the author of our own future. But it also means we can all end up in very strange and peculiar places. The most beautiful illustration of what that means comes from the badger's tale in the book *The Once and Future King*. This is a story about King Arthur. The young Arthur is educated by Merlin the Magician, who does it by converting him into animals. For instance, he becomes a hawk and Merlin teaches him about fear from the hawk.

The badger's tale is a tale about creation. At a more primitive time, in embryology, embryos were thought to be alike. All the embryos were lined up and God said, "Listen you guys, you're all sort of incomplete. I got a whole bag of stuff here. I've got wings if you feel like being flyers and I've got diggers if you want to go into the ground. I've got gills if you want to be swimmers." And so a little fish embryo came up and said, "I think I like gills." God said, "Great! You'll be a swimmer." When man came up, God said to this strange creature, "What would you like?" And he said, "Please God, I have a hunch you created us incomplete for some purposes of your own. And I will therefore stay this way all my life. If I want to dig something, I will find a tool to do it with, and if I want to fly, etc." God answers, "You alone among the creatures have divined our purpose and you will indeed have dominion over the rest of the animals because you will always be potentially in our image." I don't know that we ever complete ourselves; there is always the potential for change in the human species, as in no other species fixed by its genetic nature.

ESBJORNSON: Are there no moral and ethical limits of the creative power of the human species?

GAYLIN: Of course there are limits to creative power. I hope we haven't yet reached them.

ESBJRONSON: Should we limit the creative powers?

GAYLIN: That's a good question. I for one would be horrified by limiting creativity. I'd like to expand it.

ESBJORNSON: Is there a definable difference between respect for others and tolerance of their beliefs?

GAYLIN: I do think there is. I tend to distrust one who is too happy with conclusions about how one limits life, how one defines death, what one does with a one-pound fetus, how one handles the potential for preselection of sex of children. These are tragic dilemmas in which one must honor the fact that others on other sides have entirely different points of view. When I testified before the current President's Commission, I was against certain rules about how one should terminate life. One of the commissioners said, "Dr. Gaylin, that would make it such an agonizing decision." And my answer was, "I think certain decisions are so weighty that they ought be painful; otherwise you're not honoring them."

Following the Lecture of Clifford Grobstein

GOODFIELD: I want to refer to the case in your lecture of the investigator who went abroad to try the genetic intervention with the very best interest of his patients in mind and remind you all of one crucial fact. These regulatory protocols have two primary functions. First, they protect the human subject, the patient or the subject of the experiment, and second, they serve to protect the public's trust in the contract between the scientific community and society at large. Surely there is a question of the appropriateness of the investigator going ahead with this genetic intervention in a foreign country when the form of the protocols had not been agreed upon. The violation was not a scientific violation but it was a violation of the trust that exists between us and the scientific community. Following the recombinant DNA debates, ethical committees have come together with scientific committees to lay down protocols for particular procedures. That investigator had an obligation to get permission to do the procedure; he didn't get the permission so he went to a foreign country where that permission was not necessary. By not waiting or taking the further steps to try to convince the committee of his scientific peers that had established the protocols and going abroad, he betrayed a contractual trust and that is the issue to be addressed, not the issue whether chemotherapy or hormone replacement is just as appropriate as the genetic intervention.

GROBSTEIN: I agree with you that the prior question was, did he follow the procedures that protected both the public and the scientific community and the patient, all three? He was remiss primarily because this was a precedent-establishing situation and was bound to attract very considerable attention.

LEBACQZ: I want to take up your challenge of asking whether there are new moral principles needed and I am going to say, Yes there are. I say that as a former member of the National Commission for the Protection of Human Subjects of Research, which outlined the three ethical principles that have formed the basis for the very regulations that you think are already in place. Let me propose that we ought to have a principle that requires a covenant between those who do research and those who would be the recipients of the effects of that research for better or worse. That covenant might require some very stringent kinds of responsibility regarding the outcome of the research and a bonding between those who do it and those who receive the effects of it. This principle has been thus far left out of most of the regulations by which we operate. It is one that is consonant not only with Judeo-Christian tradition but I think with a number of other religious traditions and world perspectives as well.

GROBSTEIN: I salute you as a former member of the Commission on Human Research because certainly I think that Commission was ground-breaking both in its impact within its own field and in terms of what I refer to, namely, how we should handle this kind of matter. I think that it will always be cited as the beginning of a new trend in deliberation and policies of this kind. With respect to the principle that you suggest, I would agree with you in principle, but I find it rather difficult to imagine, for example, how in the first demonstration of recombinant DNA you would have a compact between the doers of that research and all of the people who are likely to be impacted by it.

ANFINSEN: I came here with a very simple-minded message, namely, that it's unlikely that we will ever be able to stop scientists from doing what they want to some place or another. The human animal continues over the history of man to make a lot of booboos and this will probably also continue. The rest of the people on the panel are much more likely to work out a system of controls than I am. I don't think the controls will work for a long time, but it has to be tried; so once again, standing up for the gloomy people, I say the work will go on. What we see now will be child's play when compared with what we will see ten years from now in the potential of application to human evolution. Bad things will happen. We have codes of ethics all over the place and we still have bank robberies and people cheating on construction projects. That's just the way we are.

ESBJORNSON: There are questions from the audience about such matters as the relationship between government funding and the power of government to either direct or to control these new abilities. Would you consider the government to be playing Satan?

GROBSTEIN: Our government presumably consists of people as well as traditions and restraints. To the degree that anything involves people there will be both sides of the God-to-Satan ratio. With respect to government funding and control, it is important to recognize that in our system of the funding of science, government operates largely through scientists. Particularly in the medical area, the decision-making about directions of research and about the application of new knowledge have largely been made within investigator communities, with other elements including patients, trustees, and so forth.

THOMAS: If we are to worry about individual investigators undertaking experiments of the kind that we would all agree should not be begun on human beings, there are some sanctions that can be imposed on scientists—the worst of which is to take their grants away.

ESBJORNSON: Reference was made to the Nuremberg Code and other voluntary restraints. What guarantees do we have that these Codes will be enforced and what would be the results should individuals or governments choose to defy them?

GROBSTEIN: There are no guarantees. The Nuremberg Code in particular has very wide currency and has been in some version adopted in many countries. It has been observed in the sense that some of the excesses that occurred before their adoption have not occurred since their adoption. To have the Codes seems to be a form of restraint that is acceptable and may very well have favorable effects.

ESBJORNSON: We live in a global society where there is some risk in unilateral actions. When we live in a world where the Soviet Union and other countries may have a different approach to these things, would we be able to develop any kind of global regulations in the present political climate?

GROBSTEIN: We are probably not in any position to talk to other nations until we have set our house in order in this particular area. When we clarify our own policies and put them into effect then it will be appropriate to begin serious efforts to establish some kind of international consensus. We are not the only ones concerned about these matters. With respect to *in vitro* fertilization, for example, both Australia and England are further ahead than we are in formulating effective policy. We will have to learn from them, rather than the other way around. In the case of biological warfare, where an international covenant was achieved, the problem was recognized to be of such broad scope in terms of human interest that it was possible to do something in that area that was not possible in the area of nuclear weapons control. I would hope that in this area of medical advances it might be possible to push ahead to the broad kind of consensus.

ESBJORNSON: What makes you think that an individual's wishes will be respected in genetic therapy?

GROBSTEIN: I would anticipate that an individual's wishes would be respected to the same degree, not more nor less, as in conventional therapy. The view that they're not respected at all in conventional therapy seems to me somewhat exaggerated.

ESBJORNSON: If there is an advantage in a gene such as the one related to sickle cell anemia, which also provides resistance to malaria, what will be done to correct the defect without eliminating the advantage?

GROBSTEIN: So far as I am aware, the sickle cell case is the only one in which it has been clearly demonstrated that there is an advantage that appears to be associated with the same gene. There would be no way in which one could protect the advantage while canceling out the defect of that particular gene.

ESBJORNSON: What gives you the right to turn life on and off like a light switch?

GROBSTEIN: Absolutely nothing gives me that right, and I have no wish to have that power. Nor, as a matter of fact, do I think that life can be turned on like a light switch. It may be able to be turned off like a light switch, but not on.

ESBJORNSON: Please clarify how a germ-line cell is different from a somatic cell.

GROBSTEIN: The term "germ-line" refers to that lineage

of cells derived from the original zygote, which will give rise to germ cells, namely gametes, eggs, and sperm. Once those cells have been set aside in the developing embryo, they never participate in any other activity within the embryo. They do not form any other kind of specialized cell. These are referred to as "germ-line cells" because they are germinal for the next generation. They are the precursors of the next generation. The rest of the cells in the body are somatic cells, which cannot give rise to another generation of the whole animal, although they give rise repeatedly to new generations of cells.

ESBJORNSON: As regards the concerns for gene transfer, could not the same concerns, playing God or playing Satan, have been cited for a list of other developments? Has the potential of good being used for evil not always existed?

GROBSTEIN: The questioner is entirely right. This kind of issue does come up with each new discovery. If there is anything new it is the direction in which these particular advances are moving toward far greater control over our own heredity and development than had existed in the past. The problem of playing God or playing Satan, doing good versus doing evil, has always existed so far as I know, and we will continue to wrestle with it.

ESBJORNSON: If an embryo has been diagnosed as having Down's Syndrome or some other genetic defect early in pregnancy, could it be genetically changed at that point?

GROBSTEIN: With respect to Down's Syndrome, it is a chromosomal defect rather than a defect at the level of DNA, in the sense of changes in the sequence of DNA or removal of the components of DNA. It would have to be dealt with in a very different way than the circumstances referred to with respect to sickle cell anemia. At the present time, this would probably be approached, if it were at all, by fusion of cells rather than transfer of DNA. With respect to other kinds of defects, it is theoretically possible that if they were diagnosed early in pregnancy, some form of therapy might be applied that would correct the defect in that individual. However, at the present time, very few genetic diseases have been studied with this possibility in mind, and so only very few at this time seem potentially to be correctable in these ways.

ESBJORNSON: With advances in medical science, more people with genetic defects are surviving. Will this significantly weaken the gene pool in the foreseeable future? If so, do we need gene repair for the well-being of our species?

GROBSTEIN: This is a concern that existed long before the period of recombinant DNA and gene transfer. We may be affecting the gene pool by preserving individuals who otherwise might not be able to reproduce, and having them reproduce with whatever defects they may have that account for their condition. Such a concern gave rise early in the century to an interest in eugenics to overcome this disadvantage. The circumstances under which this is occurring are changing the survival conditions for human beings. We exist within a culture and we are selected, in part, by that culture; that is, our genotypes are presumably being selected. Therefore, what may have previously been a disadvantage—to preserve certain genes that might increase the susceptibility to disease—may not be in this culture a disadvantage and in the future may not be a disadvantage.

ESBJORNSON: If it were the case that such effects on the gene pool were occurring, will we need gene repair for the well-being of our species?

GROBSTEIN: I think that it is worth considering in view of certain threats that we currently face, for example, from nuclear weapons. In the event that there was a massive release of radiation accompanying nuclear warfare, it very likely would significantly increase the genetic burden on whatever survivors there might be. Under those circumstances, techniques for gene repair might indeed be necessary, not only for well-being, but for first aid, to maintain the species in a viable state.

ESBJORNSON: To what extent is genetic engineering motivated by the potential generation of corporate profits?

GROBSTEIN: In the early stages of the present advances in genetic understanding, the potential generation of corporate profits was certainly not on the minds of the investigators involved. The initial efforts that were made in the direction of application and corporate profit-making were by small companies that were set up specifically for the purpose, and very frequently set up with the direct cooperation and even ownership of molecular geneticists who had been involved. Today, there is much more widespread interest on the part of corporations, and undoubtedly in certain areas profit-making is now a prime motivation for further advance.

Following the Lecture of June Goodfield

ESBJORNSON: In 1966 when I started to teach the Ethics in Medicine seminar at Gustavus, there was very little literature available. There has been an explosion of interest and literature in this field. An observation that I heard made today was that the ethical questions and even the theological questions that were once thought to be irrelevant and unimportant have been brought back into the public domain by the biologist. Do any of you have any similar or opposite observations about recent developments connecting science and ethics?

GAYLIN: I've never bought the idea that there's something called "bioethics." There are traditional problems that are inherent in the human condition. June brought us back to the broader issues. We tend to get philosophical when we're frightened. When things go well we forget about philosophy; we just enjoy life. What has happened is that biology, particularly current biology, has frightened us, has given us choices that we had never anticipated. Something else happened in the process. It raised the old questions. They happen to be cast now in the example of medicine. We talk about justice, autonomy, the common good, the individual's rights, family integrity, and whether the family has a place in the moral sphere. The decision about whether you should go on a kidney machine is not a medical decision. Questions about whether you should save one year of life for ten people, or ten years of life for one person, or anything like that, are all complicated moral decisions. I'm sorry if there are those of you who are somewhat disappointed that we didn't attack specific issues that you have been involved with, such as whether it really sits well morally to put an upfront hundred thousand dollar charge for a liver transplant, what is the alternative, and who is going to pay for it? Those kinds of questions have always existed. These are traditional questions to be resolved by traditional approaches to values and philosophy, but there are no experts. The lives you lead or do not lead, the deaths you experience or will not experience, and the kind of environment your children and their children will be brought into will be determined by decisions you make, or you prefer not to make, through your legislators.

GOODFIELD: I just wish to add to what Will has said. Not only are these very old questions; they are going to be very old questions a thousand years from now. These questions are never going to go away. Each generation will have to face them and sort them out and wrestle with them and deal with them.

ANFINSEN: Morals and ethics were invented by human beings for humans. We don't care very intensely about dogs being exposed to strange medical manipulation. I find myself sitting here in a slight state of amusement over the intensity of discussion about something that really applies only to our species. I think we should be more concerned with the practicalities of avoiding trouble, of avoiding threats to the human animal, rather than going into great discussions about the new moralities and relationships. That's a very negative statement but it happens to be the way I feel.

GOODFIELD: It worries me, Chris, but it's terribly humiliating and no doubt very salutary when you imply that these are "hot air" questions, and that the real questions are the questions that you have in the laboratory.

ANFINSEN: What I meant to say was that all of the metaphysical discussion we have had of various sorts is relegated pretty much to man. The main point I wanted to make was that all of these things become important only when they impinge on the human animal.

GOODFIELD: I agree with that.

ANFINSEN: I don't think we are seriously threatened yet, but we will be. It seemed to me that some of my colleagues here were concerned with the kinds of discussions that really will become critical only when we have something to shoot them at. And I'm not sure we have reached that point yet.

LEBACQZ: What you said, Dr. Anfinsen, at first seems to me in some ways to undermine, or to be directly contrary to, what June was talking about, because you were really talking about consequences as the only thing that we need to worry about. What I heard June doing was offering a sort of addendum to Will's speech. Will called upon us to ask what is distinctive or unique or wonderful about the human animal, and he offered his own solution, which had to do with potential and being unfinished, and being able to participate in our own finishing. I hear June offering us perhaps an alternative to that. And the alternative might lie in the word "caring." What may be distinctive about us is not just that we can finish ourselves or participate in our completion, not just that we can worry about consequences, but that we have the capacity to care for each other. I thought that your holding together of rigor and of care and compassion is a particularly nice conjunction. It does seem to me to offer us a different agenda for the human animal than the one that Will was giving us earlier.

GAYLIN: I take loving and caring for granted.

LEBACQZ: Oh, never, never do that. What a lovely life you must have lived! June, you have urged us to be loving and to be caring, though you indicated that there might be institutions that would either help create or destroy loving and caring. You really have not given us a sense of how we create caring human beings, and I'm not personally convinced that it comes about through the urging of people to do it.

GOODFIELD: I agree. I refer again to eradicating smallpox in Bangladesh. I know that for some people that experience of going out there and doing it in that country of seventy-five million people was an experience that actually generated caring. Some hardboiled reductionist scientists thought they could go right out there and clear out smallpox and walk roughshod over everybody in order to do it. They found they couldn't walk roughshod over everybody. It may be through experiences like that that you become caring. There are a whole variety of ways, but I can't answer that question.

GROBSTEIN: As I listened to Will, I certainly agreed with what he began with. But then things got a little confusing for me. Particularly since all of this is done, at least partly, in the name of biology. And biology, looked at in terms of all of its aspects, is not a particularly caring science. I think it is a gentle science, as compared with physics, which deals with huge quantities, enormous distances, enormous energy, and the like. Biology tends to deal with miniscule things. Life can only survive within a fairly narrow range of temperatures and impacts of energy and so on. But caring . . . not in biology. Certainly caring comes later than that. It comes in what I would have to call the human side. Biology has its own sphere and it ranges fairly widely in terms of levels of organization from macro-molecules to almost the kind of phenomena we talked about in human caring. But it stops somewhat short of that. So I think maybe when you were pointing out, Bob, that this thing has happened in the name of bioethics, I was inclined to feel that what we are really talking about is not bioethics, we are still talking about medical ethics.

ESBJORNSON: I like that definition of sin used by June in her lecture, "the heart turned in upon itself." What is it that uncurves us, so that we do turn outward?

GAYLIN: That's a biased question. I assume that it's natural to be caring and loving. If we know anything from our development, we know that we cannot survive without a network of supporting individuals. We are born grotesquely helpless, and we remain helpless for a prolonged period of time. Could such a creature have survived at all for twelve or fifteen years of being impotent and helpless if there weren't built into the organism some caring and loving features to enhance or secure that survival? The problem is that we have developed cultural institutions to numb and destroy that natural compassion. The question is, what have we done to our culture that has caused it to take naturally caring, empathetic, and loving individuals and squeeze them out of all semblance to what they were intended to be?

LEBACQZ: I want to answer the question about how we create caring human beings. I agree in part with you, Will, but not totally. I do think that humans are meant to have the capacity for caring, and that we are probably born with it. I also, perhaps because of my Lutheran background, am a great believer in the Fall, and I do not think that human nature as it exists in this world tends to live up to all of those wonderful capacities. Your mention of the institutions that can destroy our capacity for caring is only one example of the "Fallenness" of our world. What do we do in the face of that, other than to try to make our institutions better, which I heartily support? I would say that the stories that we tell our children and ourselves are very crucial in determining the kinds of people that we become. The stories of Winnie the Pooh are among my favorite stories. If you do not raise your children on *Winnie the Pooh*, then you are doing something wrong. Many of the young people that I have met at this conference must have been raised on the stories of Winnie the Pooh. I want to conclude my time with you by saying thank you to all of you, but particularly to the young people here for the gift of graciousness and care and thoughtfulness and intelligence and warmth that you have given me, and in return I give you a gift. May you have a parable a day to keep the monster away!

GROBSTEIN: I take away a very deep impression of having had the opportunity to interact not only with members of the panel, but with a community which, it seems to me, is among the most favorable that I've ever experienced. I carry away an impression of young people who are deeply dedicated, from whom we can expect to hear in the future.